About Island Press

Since 1984, the nonprofit organization Island Press has been stimulating, shaping, and communicating ideas that are essential for solving environmental problems worldwide. With more than 1,000 titles in print and some 30 new releases each year, we are the nation's leading publisher on environmental issues. We identify innovative thinkers and emerging trends in the environmental field. We work with world-renowned experts and authors to develop cross-disciplinary solutions to environmental challenges.

Island Press designs and executes educational campaigns, in conjunction with our authors, to communicate their critical messages in print, in person, and online using the latest technologies, innovative programs, and the media. Our goal is to reach targeted audiences—scientists, policy makers, environmental advocates, urban planners, the media, and concerned citizens—with information that can be used to create the framework for long-term ecological health and human well-being.

Island Press gratefully acknowledges major support from The Bobolink Foundation, Caldera Foundation, The Curtis and Edith Munson Foundation, The Forrest C. and Frances H. Lattner Foundation, The JPB Foundation, The Kresge Foundation, The Summit Charitable Foundation, Inc., and many other generous organizations and individuals.

The opinions expressed in this book are those of the author(s) and do not necessarily reflect the views of our supporters.

STRUCTURES OF COASTAL RESILIENCE

STRUCTURES
OF COASTAL
RESILIENCE

Catherine Seavitt Nordenson,
Guy Nordenson, and
Julia Chapman

ISLANDPRESS Washington | Covelo | London

Structures of Coastal Resilience

ISLAND PRESS is a trademark of the Center for Resource Economics.

Keywords: Adaptation, adaptive management, climate change, Coastal Storm Risk Management (CSRM), Federal Emergency Management Agency (FEMA), flood risk, floodplain management, National Flood Insurance Program (NFIP), National Oceanic and Atmospheric Administration (NOAA), nonstructural flood risk management measures, sea level rise, storm surge, United States Army Corps of Engineers (USACE), United States Geological Survey (USGS), vulnerability, water quality, wetland

Library of Congress Control Number: 2017958675

All Island Press books are printed on environmentally responsible materials.

Manufactured in the United States of America
10 9 8 7 6 5 4 3 2 1

CONTENTS

Foreword

Michael Kimmelman

ARCHITECTURE CRITIC, *THE NEW YORK TIMES*

A while back, having dragged a borrowed kayak out of the Los Angeles River, I called a friend in West Hollywood whom I had arranged to meet. I told him I was running late. My trip had taken longer than I'd anticipated, and I was soaking wet in a remote stretch of woods. He fell silent for a long time.

"What river?" he finally said.

The Los Angeles River, around which the city first evolved, can be a dim concept even for lifelong Angelenos. What may vaguely come to mind is the concrete basin from *Terminator 2*, glimpsed from the Sixth Street Bridge: the area, downtown, that the United States Army Corps of Engineers canalized many decades ago after a series of floods socked the city. Much of the rest of the river, including the part where I kayaked, has long been obscured by rail tracks and highways—the neglected backyard in poor East Los Angeles and other neighborhoods.

This is because the Army Corps' techno-bureaucratic attitude for much of the twentieth century, like the attitude of most public officials, was that water was an enemy. It needed to be subdued, sequestered from valuable real estate. Hydrologic solutions demanded hard structures—revetments, bulkheads, and levees. There was next to no concern for how cities and regions might live *with* nature rather than fight against it; how the destruction of coastline, marshes, and floodplains might come back to bite us; and least of all, little or no thought given to design or aesthetics (landscape architecture seemed frilly and expensive). As a consequence, in a city such as Los Angeles, millions of people ended up cut off from the river, many from jobs, the urban fabric sliced up in ways that isolated vulnerable neighborhoods.

There is no bigger challenge today than the management of coastal ecologies. With climate change, this challenge has begun to take on an existential dimension, threatening the whole global economy and the stability of nations. Hurricanes, heat waves, and floods that used to be rarities are becoming the new normal. The majority of people on Earth today live on or near coasts, a number swiftly escalating in this first urban century in human history, with coastal population densities twice the world average. Most of our largest cities are coastal megalopolises, and their growth (from Houston to the Pearl River Delta, Miami to Jakarta) has involved runaway development, with vast, critical swathes of mangrove, wetlands, prairies, and forests, which mitigate the impact of storms and rising seas, ripped out to make way for miles and miles of concrete and asphalt.

At the heart of this book is the role design needs to play in devising new approaches to coastal resilience. As we've witnessed over and over, we can't simply continue to battle nature with walls and gates. They won't suffice, and they're often counterproductive. The book rethinks "structural" solutions to mean more than things built out of concrete and steel. There also need to be structures in place for dealing with politics and people. The book embraces an emerging paradigm shift toward risk management, encouraging climate scientists and engineers to team up with designers, planners, and community leaders. Its goal is to replace dated concepts of flood control with strategies for controlled flooding: hard infrastructure, which is never infallible, with a mix of hard and soft tactics that can produce all sorts of benefits aside from keeping people's feet dry and property safe.

This is in keeping with a broader shift in progressive thinking that has taken place during the twenty-first century. In cities such as Rotterdam, Madrid, Seoul, and New York, waterfronts are being recuperated, turned into parks, and treated as assets, not obstacles. The movement to include ecological and landscape design into strategies for coastal resilience has run headlong into entrenched politics, neoliberal economics, and a very common human desire to live wherever we want, never mind the consequences, but it has been gaining a voice in America at least since Hurricane Katrina. Now it has penetrated thinking at many architecture schools and government agencies, including the Army Corps, which remains a cumbersome and often counterproductive player but has shown itself to be more open to evolving ideas about coastal protection and water management, even at sites such as the Los Angeles River, where social and economic logic dictates that neighborhood revitalization go hand in hand with wetland restoration and new infrastructure for capturing water. This is because the huge, drought-plagued city can no longer afford simply to dump billions of gallons of water from the river into the ocean. Frank Gehry is the latest architect to join in this effort, a sign of recognition that architecture and design are integral to progressive resilience planning.

Famously, the Dutch have been promoting an integrated approach to water management for a long while. With much of the Netherlands below sea level and sinking, living with water has been a daily matter of survival for as long as the country has been around. The Dutch tried building massive fortifications to stop the North Sea. In 1916, a storm overwhelmed the Dutch coastline, inaugurating a spate of construction that failed to hold back the water in 1953 when an overnight storm killed more than 1,800 people. The Dutch still call it The Disaster. They redoubled national efforts, inaugurating the Delta Works project that dammed two major waterways and produced the Maeslantkering, the giant sea gate, completed in 1997, a spectacular feat of architectural engineering that keeps open the immense canal that services the port of Rotterdam.

But dikes and gates still weren't enough, they realized, because the North Sea wasn't the only threat. With climate change, the rivers that flowed into the country were swelling and posed an even greater challenge. So the Dutch in recent years undertook a program called Room for the River, which found ways to turn polders into retention ponds and allow retention ponds to double as recreation sites, dikes to double as parks, and industrial riverfront to morph into leafy new neighborhoods, business incubators, and tourist attractions. Environmental and social resilience go hand in hand, the Dutch maintain, and protection from water should also improve neighborhoods, promote public health, and make cities richer, more attractive, and more equitable— because those cities will be more capable of facing the inevitable stresses climate change imposes on society. That's what resilience actually means.

This book examines North American sites such as Jamaica Bay in New York, where the threat of rising seas to an already fragile ecosystem is complicated by a legacy of public housing from the Robert Moses days and the existence of critical city infrastructure, including John F. Kennedy International Airport. Proposals for protection involve a mix of natural and nature-based solutions that promote the existing marsh ecology and also support long-standing communities, for whom the endangered bay is home. In the long run, what's economical is also what's environmentally and politically sustainable, the book argues. And this is where design plays a key role.

To live with the challenges of the new urban century requires devising cities where people *want* to live, affordable places that are healthy and efficient, that promote diversity, attract new businesses, and look beautiful. Design is not a frill. It is necessary—good design, anyway. Pennsylvania Station, in New York, is the hemisphere's busiest and possibly most horrendous transit hub, a dangerous firetrap and shameful underground rat's maze, serving some 650,000 people a day, roughly equivalent to the population of Boston. Those passengers suffer an aging rail system that constantly breaks down, with a pair of century-old passenger rail tunnels under the Hudson River, strained to capacity, whose days are numbered. When one of those tunnels fails, the ripple effects will register on the national gross domestic product.

But new tunnels aren't enough. New tunnels without an improved Penn Station—a station that's inviting, safe, and even inspired, like Grand Central Terminal—is tantamount to buying a fancy garden hose without swapping out the rusty little bucket the water pours into. What's needed is a place worthy of New York, attuned to the city's aspirations and democratic ideals. This requires architecture and fresh ideas about public space. For millions of riders, it's a basic matter of dignity and equity.

Design married to improved engineering, social justice, and economic growth. That's a better strategy for the new century. At the waterfront and elsewhere, it's the only way forward.

Acknowledgments

The authors are happy to have the opportunity to thank the many individuals and institutions who supported the successful completion of this book. The Rockefeller Foundation deserves special mention, as it generously supported the *Structures of Coastal Resilience* (SCR) design initiative with a research grant in 2013, allowing us to gather a coalition of universities to work closely with the United States Army Corps of Engineers (USACE) North Atlantic District, exploring new strategies for coastal protection during the 2 years after the landfall of Hurricane Sandy on the Atlantic seaboard. Nancy Kete and Samuel Carter, former managing directors of the Resilience Team at the Rockefeller Foundation, provided an unerring vision for the project's success. Chief Joseph R. Vietri and deputy director Roselle Henn Stern, both of the USACE North Atlantic Division's National Planning Center for Coastal Storm Risk Management, welcomed the collaborative design approach initiated by SCR. Peter Weppler, chief of the Environmental Analysis Branch, and Lisa Baron, project manager of Civil Works, both of the USACE New York District Office, provided invaluable suggestions and support throughout the process. We are indebted to our SCR science team at Princeton University for their innovative work on sea level rise and probabilistic storm modeling for this project, with a special acknowledgment to Ning Lin, Michael Oppenheimer, Talea Mayo, and Christopher Little. We are also grateful to the SCR collaborative design teams at the four participating universities, led by principal investigators Michael Van Valkenburgh and Rosetta Elkin at Harvard University, Catherine Seavitt at City College of New York, Paul Lewis at Princeton University, and Anuradha Mathur and Dilip da Cunha at the University of Pennsylvania. Julia Chapman's role as project manager of SCR was invaluable, and her diligence and insight were instrumental to the success of the project, and indeed the completion of this book. Each SCR design team was made up of invaluable individuals, and we are grateful for their talents and contributions to the project, with special thanks to Michael Tantala, Elizabeth Hodges, Enrique Ramirez, Tess McNamara, Emma Benintende, Kjirsten Alexander, Danae Alessi, Eli Sands, Michael Luegering, Marissa Angell, Michalis Piroccas, Anna Knoell, Kevin Hayes, Caitlin Squier-Roper, Jamee Kominsky, Graham Laird Prentice, and Matthew Weiner.

We would like to acknowledge the Museum of Modern Art (MoMA), its director Glenn Lowry, and its curator Barry Bergdoll, who previously served as director of the Department of Architecture and Design from 2009 to 2013. Barry created a radical new workshop program at the MoMA, with its first iteration in 2009–2010 titled "Rising Currents: Projects for New York's Waterfront." Also funded by the Rockefeller Foundation under the leadership of former president Judith Rodin and former vice president Darren Walker (now president of the Ford Foundation), this successful workshop and exhibition was initiated by the design research presented in our book *On the Water: Palisade Bay* (Hatje Cantz/MoMA, 2010), co-authored with Adam Yarinsky. We remain especially grateful to the Fellows of the American Institute of Architects (FAIA) and Latrobe Prize jurors Daniel S. Friedman, FAIA, chair; Frank E. Lucas, FAIA, chancellor; Martin Fischer, PhD; Leon R. Glicksman PhD; Frances Halsband; and James Timberlake, FAIA, for awarding us the biannual Benjamin Henry Latrobe Prize in 2007 for the proposal "On the Water: A Model for the Future New York and New Jersey Upper Bay." This research project became the aforementioned book and initiated our ongoing work on coastal resiliency in light of climate change and sea level rise.

We acknowledge the ongoing intellectual support of our home academic institutions, the City College of New York and Princeton University. Thanks are due to the interim dean of the Bernard and Anne Spitzer School of Architecture at City College, Gordon Gebert, and to the former dean of the Princeton University School of Architecture, Stan Allen, as well as the generous and capable administrative staff at both institutions. We are also grateful for the stimulating academic environment provided by our talented colleagues, with particular thanks to Michael Oppenheimer, Albert G. Milbank Professor of Geosciences and International Affairs in the Woodrow Wilson School of Public and International Affairs and the Department of Geosciences at Princeton University, who has enthusiastically supported our coastal design initiatives from the very beginning.

We are indebted to the generosity of our many design colleagues who contributed their work to this book, with special thanks to Paul Lewis, Marc Tsurumaki, and David J. Lewis of LTL Architects; Kate Orff of SCAPE; Adam

Yarinsky, Stephen Cassell, and Kim Yao of Architecture Research Office; Mimi Hoang and Eric Bunge of nARCHITECTS; Matthew Baird of Matthew Baird Architects; and Susannah Drake of DLANDstudio. We are also grateful to the many artists—and their estates and archives—whose creative work and unique perspectives presented herein continue to deeply inspire us.

Two individuals deserve a very special note of gratitude for their contributions to this book. Michael Kimmelman, architecture critic of *The New York Times*, provided the thoughtful foreword, and we are indebted for his continuing interest in and championing of the work that designers envision for the urban realm. Jeffrey P. Hebert, the vice-president for adaptation and resilience, The Water Institute of the Gulf, wrote the insightful and timely afterword. His ongoing work on the ground is the realization, through advocacy and policy, of the design visions proposed by this book.

We are particularly grateful to Island Press and our editor Courtney Lix, whose support of this project from its initiation as a simple prospectus was unflagging and whose insights, edits, and advice regarding its advancement were consistently on the mark. Courtney's comments, suggestions, and provocations led to a significant reshaping and clarification of the manuscript and its arguments. Our thanks to Island Press extend to vice president and executive editor Heather Boyer and to the copyediting and production teams for their exceptional work throughout the book's editing and design process.

Finally, our greatest acknowledgment is to our families, and especially to two boys, Sébastien and Pierre Nordenson, for their patience, curiosity, and interest in this enduring project that has encompassed their entire lives. Our work, ultimately, is for them.

Chapter 1

Designing for Coastal Resiliency

For months and even years after a hurricane, images of water in the city are haunting. In October 2012, Hurricane Sandy transformed New York City's subway tunnels into rushing underground canals. Photographs captured fields of taxicabs floating in the water, coastal homes collapsing into the ocean, and floodwaters cresting over the stoops of nineteenth-century townhouses. More than a decade after Hurricane Katrina struck the Gulf Coast and New Orleans in 2005, the aerial surveys of submerged city blocks captured by helicopter overflights remain ingrained in national memory. Photographs on the ground revealed New Orleans residents trudging through chest-deep muddy water and navigating city streets and highways in boats and makeshift rafts. More recently, the trio of massive superstorms of the 2017 hurricane season—Harvey, Irma, and Maria—significantly affected the Texas Gulf Coast and Houston, the Virgin Islands and the Florida Keys, and the entire island of Puerto Rico. Indeed, the hurricanes of 2017 again revealed not only the vulnerabilities of coastal cities but the painful inequities of fragile social and infrastructural systems. Once again, images capture the incomprehensible hardship and loss induced by hurricane-force winds, torrential rainstorms, and massive storm surge flooding. These images are also uncanny; the juxtaposition of water with urban structures is unfamiliar and disquieting.

By contrast, some of the most beautiful portraits of Venice capture the city during the *acqua alta*, the exceptionally high tides of the Adriatic Sea that bring water into the plazas and streets of this Italian city. For centuries, Venice has flooded on a regular basis. Architectural historian Manfredo Tafuri deemed Venice the first modern metropolis, in part because it was the first city to be built without medieval walls. Instead, the shallow lagoon surrounding Venice secured the city from foreign intruders with its mudflats, hidden channels, tides, and currents. The sea provided a defense system for the Venetians that perimeter walls could not. The safety and security of the city depended on bringing the sea in.[1]

Systems of levees, seawalls, and barriers built to defend many coastal cities from flooding diverge from this Venetian approach. Yet as recent flood events have shown, these hard infrastructural defenses are not infallible. Indeed, during massive rain events, coastal defenses—designed to prevent surge flooding from the ocean—prove useless and may even exacerbate inland flood damage. When flood infrastructure fails, disaster often ensues. In addition, sea level rise, one of the many tangible impacts of global climate change, has presented the imperative of thinking differently about coastal flooding. A warming climate leads to rising water levels, with a massive impact on coastal cities. As oceans warm with the increase in global temperature, seawater expands within the ocean basin, causing a rise in water levels. In addition, the warming climate is causing the significant melting of ice over land at higher latitudes, which then adds more water to the ocean. One evident effect of climate change is that hurricanes are causing frequent and costly damage to areas that were historically at low risk of routine flooding. With sea level rise, flooding is likely to occur more frequently, even during average storms or very high tides. The urban neighborhoods, industrial sites, and tourist destinations that have developed in floodplains have often thrived because of their proximity to the water, but rarely have design and planning been used to mitigate potential flood damage in these vulnerable places. The result has been repeated catastrophic damage from strong storms, as well as increased chronic tidal inundation, often called "nuisance flooding." Coastal cities must adapt to keep pace with the transforming dynamic between ocean, land, and climate. Sea level rise and the increased risk of storm surge demand new solutions to the management of coastal flood hazards. Attempting to keep the water out of these communities is no longer a realistic strategy. Rather, creative planning and design might envision ways to allow water to enter these neighborhoods while reducing the risk of damage to property, livelihood, and health.

Coastal resilience necessitates not only new infrastructural strategies but also a fundamental transformation in understanding how cities might relate to the water around them. Sea level rise and storm surge may be interpreted not as threats to urban life but as opportunities for re-envisioning new ways of living along the coast. Climate change might trigger new urban paradigms; it could be the inspiration for imagining the transformation of cities into a new condition that is not only resilient to climate hazards but also greener, healthier, and more equitable.

This book seeks to offer information of value to the policymakers, engineers, scientists, planners, architects, and landscape architects who play an enormously important role in rethinking and visualizing a resilient built environment. Developing a comprehensive and integrated approach to coastal resilience will require a greater emphasis on methods of design thinking than has traditionally been allotted to this category of work at the waterfront. Imagining alternative futures for coastal cities and landscapes will require new paradigms of innovative analysis, synthesis, and design—productive methods and creative strategies for considering the complex parameters of both ecological systems and uncertain climate futures. Resilience discourse and disaster management are typically dominated by numerical analysis. Yet the value of visual imagination and spatialized system thinking in this arena cannot be underestimated.

Todd Shallat's authoritative analysis of the early history of the U.S. Army Corps of Engineers (USACE), *Structures in the Stream: Water, Science, and the Rise of the US Army Corps of Engineers*, illustrates the ordering of the landscape and its waterways not merely as the result of physical structures but as part of an ongoing process of defining and shaping institutional structures.[2] This book draws on Shallat's holistic interpretation of "structures," emphasizing the connections between physical structures—dams, dunes, seawalls, levees, revetments—and nonphysical structures—laws, policies, institutions, bureaucracies. These two kinds of structures are intertwined and must be understood in tandem. Structures of coastal resilience therefore are broad in both scale and scope; they encompass not only physical proposals but also the conceptual and theoretical constructs that define the complex systems at play along the coastline, including both opportunities and hazards.

The term *resilience* also carries nuanced associations. In the last decade, *resilience* and *adaptation* have entered the vocabulary of city, state, and federal agencies. The pervasiveness of these terms illustrates not only rising awareness that climate change will bring increased and irreversible environmental risks and disturbances but also an increase in the cumulative efforts to take action. At the local level, coastal cities and counties as varied as Portsmouth, New Hampshire; Plymouth, Massachusetts; and Lafourche Parish, Louisiana are developing resiliency and adaptation plans to cope with coastal flood risk, wetland restoration, and emergency management. Federal agencies, including the National Oceanic and Atmospheric Administration (NOAA), the U.S. Geological Survey (USGS), the U.S. Department of Housing and Urban Development (HUD), the U.S. Fish and Wildlife Service (FWS), and the USACE have all devoted resources to resilience efforts, especially after Hurricane Sandy in 2012. In his 2013 Climate Action Plan, President Barack Obama emphasized the need to prepare communities for climate impacts and increase the resilience of buildings and infrastructure. A 2015 Executive Order requires that new federal investments in floodplains manage risk at a higher standard.

Although the term *resilience* is rapidly gaining new applications in diverse fields including economics, sociology, and education, its origins are primarily in engineering, psychology, and ecology.[3] And although *resilience* is now a buzzword in contemporary political and planning circles, these multiple origins suggest that *resilience* carries several different definitions. Engineering and ecological resilience, in particular, have distinct meanings rooted in their respective fields. The development of ecological resilience theory began in the 1960s. By the 1970s, ecologists pioneered ideas surrounding "alternate states" and "regime shifts," arguing that an ecological system does not maintain a steady-state condition, reacting only to disturbances and then returning to that steady state. Rather, ecosystems are constantly in flux, shifting states or regimes though natural and human-made disturbances.[4] In his 1996 essay "Engineering Resilience versus Ecological Resilience," ecologist C. S. Holling distinguished between two types of resilience: After an event or interruption, engineering resilience pursues the return to a steady-state condition, whereas ecological resilience accepts major regime shifts in the system in order to survive or maintain function.[5] The survival of complex ecological systems depends on the acceptance of, and adaptation to, a fundamentally changed environment.

Ecological resilience theory suggests that change in a system is both inevitable and critical. Coasts are dynamic entities, changing shape and form in reaction to powerful forces of water and wind. Cities, too, are always in flux. Gentrification, blight, growth, decay, renewal, and revitalization transform neighborhoods over years and decades. An understanding of resilience as *ecological* resilience capitalizes on the dynamic processes that make coastal cities vibrant places to live and work. In this book, coastal resilience does not mean restoring coastal communities to preflood conditions or creating a steady state; rather, it means building dynamic systems that can transform, change, and evolve during and after a flood. Ecological resilience readily encompasses climate adaptation, the process of planning for expected yet unknown future environmental conditions.

But coastal resilience goes beyond accepting change; it actively encourages transformation, creating *places* that embody shared visions for the city. In coastal cities, citizens should be able to see, hear, and smell the water. Rivers, bays, and oceans are critical assets for urban life. Coastal resilience encourages infrastructure that exploits rather than disrupts this connection to water. In some cases, this means expanding notions of infrastructure beyond hard engineered structures to encompass what has come to be known as "green" or "soft" infrastructure: wetlands, islands, forests, parkland, and vegetation but also floodplain management and planning. Incorporating geological, ecological, and biological systems into coastal infrastructure is important not only for mitigating flood risk and preserving vulnerable ecosystems but also for creating novel urban spaces that can withstand change. Coastal resilience transforms the threats of rising seas and more frequent surges into opportunities for building stronger cities.

This understanding of resilience builds on sustainability, another key term of an environmentally conscious age. From building to fishing to farming to manufacturing, sustainable practices aim to minimize environmental impact, reducing natural resource consumption, greenhouse gas emissions, and other pressures on human and natural ecosystems. With its root word *sustain*, sustainability assumes that the planet might be sustained in a steady-state condition. But with climate change, the nonstationarity of climate, weather, and severe events promises a tumultuous future in which the status quo cannot be maintained, and indeed must adapt. Sustainability also tends to be parceled. A single building or consumer product may be deemed "sustainable" by accrediting parties or even marketing agencies. However, resilience is not so easily compartmentalized. Communities and cities build resilience through collaborative efforts and comprehensive initiatives, considering the relationships between systems, functions, and populations.

Similarly, design thinking, which this book emphasizes, encourages consideration of the relationships between systems, functions, and populations. Multiple scales must be considered, and multiple time horizons. Design thinking requires imagining a different kind of city in a postdisaster context, one that may look and work differently than before. For example, notions of resiliency that prompted a shift toward design thinking in New York City may be traced not only to the impacts of Hurricanes Irene and Sandy on the region in 2011 and 2012 but also to the earlier national crisis of Hurricane Katrina in 2005, when it became clear that some of the greatest future challenges for urban planning and natural disaster mitigation would be compounded by climate change. The cultural heart of the New York region has been rapidly extending beyond the island of Manhattan into the outer boroughs and New Jersey. The New York and New Jersey Upper Harbor is a new center of gravity, pulling development and design attention toward the Manhattan, Brooklyn, Staten Island, Hoboken, and Jersey City waterfronts. This local interest in the harbor—for both its urban potential and its environmental risks—was compounded by the 2005 hurricane season. When Hurricanes Katrina and Rita struck the Gulf Coast in August and September of 2005, sea level rise, storm surge, and the disappearance of coastal wetlands garnered national attention. At that time, the possibility of a hurricane striking New York City was not at the forefront of municipal planning, although the region falls within the zone of possible serious storms. And throughout the twentieth century, several storms had caused severe damage to the New Jersey and Long Island coasts. This context incited not only concern for the impact of climate change on the region but also ideas for how design might simultaneously mitigate risk and transform the metropolitan area.

Interest in resilient design in New York was heightened by the aftermath of Hurricane Sandy in 2012. Coastal resilience gained sudden urgency. Projects previously received by the public as speculative gained renewed interest as critical and feasible. Moreover, projects that presented a systemic approach to resilience that would work with natural processes, rather than against them, gained traction. This significant paradigm shift must continue in the devastating aftermath of Hurricanes Harvey, Irma, and Maria in 2017. Innovative design work reconsiders traditional flood control practices, exploring the potential for a Venetian-influenced adaptation of "controlled flooding." It embraces water currents and tidal exchange as part of the design and construction process, not as obstacles to a new vision of urbanism. It prioritizes ecologically sound waterways as part of a resilient coast. Vegetation—in the form of designed wetlands, mangrove forests, and maritime forests—attenuates waves and reduces the impact of storms on coastal landscapes. These softer infrastructural strategies and systems are appealing every day of the year, not only during times of flood or surge, and are adaptable to uncertain climate futures. They also highlight the power of design and planning to transform the relationship of a place to the dynamic systems of water that interact with it. Much of what design thinking engenders in future communities will rely on creative techniques, but Venice remains a reminder that visionary humans have adapted to the unpredictable influxes of natural systems for centuries.

The design projects presented in this book provide tangible visions of what urban coastal adaptation could look like, from the Palisade Bay research project, to the creative New York City waterfront projects it engendered at the Museum of Modern Art's workshop and exhibition *Rising Currents*. The research project *Structures of Coastal Resilience*, developed by four universities working in parallel with the USACE, examined the coastal embayments affected by Hurricane Sandy at Norfolk, Virginia; Atlantic City, New Jersey; Jamaica Bay, New York; and Narragansett Bay, Rhode Island. Beyond the North Atlantic coast of the United States, collaborative design studies addressing coastal protection and wetland restoration through sediment diversions along the Mississippi River and its delta are presented, along with new ideas for improving water capture, storage, and release at the Yangtze River Delta in China based on historic agricultural techniques. These projects illustrate an approach for creating storm-resilient landscapes by integrating engineered ecologies with traditional storm protection infrastructures. The design of layered systems allows for controlled flooding but also reduces the risk of damage from storm surge, enhances ecologies, and improves the quality of daily life for local residents.

This book's approach to the planning and design of resilient coastal communities draws on innovative research, partnerships, and projects conducted over the past decade—local, national, and international in scope. Although each coastal resilience project addresses a specific site and its parameters, they all share a common method and embrace design thinking. All the projects are transdisciplinary in scope, involving design thinkers and scientists from many fields, as well as policymakers and government organizations. Working comprehensively and holistically across urban, ecological, and engineered systems, these projects provide a humanist impulse to a field typically dominated by technical expertise, bureaucratic regulation, and hard data. Each initiative demonstrates the capability of resilient design to build robust and adaptive cities through urban planning and landscape design strategies. And each illustrates the role that architects, landscape architects, and planners can play in addressing climate change at a large scale.

The organization of this book follows the working method of many of these projects, from site analysis to design to evaluation. Throughout, it explores the productive exchange and interaction between the design process and scientific method. The second chapter examines the ways innovative designers have begun to interpret coastal sites by combining the representational methods of art and architecture (plan and section drawings, collage, scale models) with tools of coastal engineering and geospatial analysis (wave flumes, hydrodynamic models, and digital elevation topographic and bathymetric data). Research into these ideas was initiated with the Palisade Bay project analyzing the Upper Harbor of New York and New Jersey in 2007–2010 and continued with the *Structures of Coastal Resilience* projects of 2013–2015. Pushing the conventional boundaries of representational techniques not only reveals the hidden conditions of familiar sites, it also generates inspiration for new design interventions. Visualizing dynamic processes so that they become both accessible and easily understood is critical to communicating how climate change will affect coastal environments. Representation and visualization is the first step toward creating the novel designs and plans that will reflect the dynamic futures of coastal communities.

The third chapter moves beyond visual representation into examples of the collaborative process of interdisciplinary design thinking applied to the projected futures of a resilient coastal condition. It explores coastal projects at a range of scales, from small landscape interventions to urban design initiatives to regional planning efforts targeting an entire river delta. The chapter examines how two kinds of structures—both physical infrastructure and the processes by which institutions operate—might complement each other in the context of planning for coastal resilience. Each project imaginatively combines the three key principles of *attenuation*, *protection*, and *planning*. *Attenuation* and the dissipation of wave energy offshore reduces the demands on barriers and levees with wetlands, breakwaters, islands, or offshore structures; *protection* of the built environment with both flood mitigation structures and code requirements anticipates inevitable flooding; and *planning* for controlled floods through urban and landscape floodplain management and design manages risk effectively. Taken together, these principles may be creatively applied to a layered system of protection, comprising natural, nature-based, structural, and nonstructural measures that can effectively produce resilient and adaptive solutions for coastal communities while accommodating varying levels of flood risk given future climate scenarios.

The fourth chapter further develops the critical combination of strategies from both the design and science disciplines to map potential coastal flooding and visualize how design interventions might mitigate this flooding. It demonstrates the importance of evaluating design proposals through the lens of precise climate science and coastal flood assessments, with attention to the dynamic conditions that surround resilience and adaptation. Coastal hazard assessments, which evaluate storm surge and sea level rise probabilistically throughout the next century, provide information about how the risk of flood events will be transformed over time given various scenarios of climate change. These hazard assessments can be represented spatially and graphically, through maps and matrices that provide designers and planners with information about the changing risks of flood inundation in a specific study region. Drawing from the performance-based design method used in fire, wind, and seismic engineering, this chapter proposes a dynamic performance-based design approach for coastal flood hazards. This approach offers planners, designers, and stakeholders a means to understand the changing levels of hazard and risk over the lifetime of a project and identifies how design might respond to frequent but mild flooding as well as to infrequent but severe storm events.

Climate change is a defining challenge of the twenty-first century. It will affect critically important issues: income inequality, education, healthcare, economic growth, and energy. At the same time, the urgency and severity of climate threats offer great opportunities for radical and social interventions developed and executed with both design imagination and scientific rigor. The fifth chapter concludes with a reflection on the often-evoked time scale of one hundred years by looking not only forward to 2100 but back in time, reflecting on the impacts and outcomes of coastal planning projects from the early twentieth century. Much can be learned from the unintended consequences of well-intentioned human design interventions on the coast. And although many strategies have had negative ecological and social consequences, much may be gleaned from examining what were in some cases brilliant foundational principles: the connection of the environment with human health, the civic scale of urban adaptation, and the desire to create equitable and affordable urban housing. As in Venice, where letting the sea in was an intentional act of security, twenty-second-century coastal cities might expect and even celebrate a new watery urbanism, an ecological and human connection to aqueous systems that supports an equitable, robust urbanism. Keeping water out may in fact provoke destruction, but letting it in could lead to a rich, adaptive, and vibrant urban future.

Chapter 2

Visualizing the Coast

To represent the coast means reckoning with the dynamism of water. To draw water, to capture particles in constant motion with static lines, tones, textures, or shades, seems impossible. From the regular rhythms of ocean tides and deltaic flows to sporadic storm surges and tumultuous waves, water moves across vast territories. And from the slow and powerful geological processes of erosion to the frightful rapidness of a flash flood, water carves and inundates, shaping the land beneath and around it. Never static, the coast is always in flux.

American artist Claes Oldenburg beautifully captured this instability of coastal conditions with his 1965 collage "Notebook Page: Spills of Nail Polish, 'Jamaica Bay.'" (See Color Plate 1.) Glossy blobs of nail polish lacquer evoke the marsh islands of New York's Jamaica Bay in Brooklyn and Queens. Thick and bubbly, with unwieldy edges, the dark red and pink islands appear to be caught in a moment of stasis, just about to spread across the paper. Indeed, as intertidal marshlands, the real islands' perimeters continually shift and morph. Oldenburg cleverly writes that his work is a "proposal to beautify the approach to Kennedy Airport by coloring the many little islands of Jamaica Bay,"[1] revealing that the Jamaica Bay marsh islands already function aesthetically when viewed from above by travelers on their descent into New York City's John F. Kennedy International Airport.

Oldenburg's expanding blobs of nail polish capture flux; too often, the coastline is understood, represented, and maintained as a hard static line. When a coast is imagined and recorded as a line, it is also often constructed as such, from engineered seawalls to urban waterfront boardwalks. But sea level rise and an increasing risk of inundation from future storm surges challenge any fixed demarcation of coastal edges. Climate change calls into question not only the physical forms of coastal protection but also the means and methods through which the coast is understood and represented. To understand how climate change will affect a coastal site, concepts and categories must extend beyond the established oppositions of land and water, wet and dry, constructed and natural. This new paradigm requires inventive methods of analysis, interpretation, and representation.

At a moment when cities are rapidly rediscovering, revitalizing, and renewing urban coastal areas, this ability to re-envision the coast is increasingly critical. Waterfronts were historically sites of industrial production and distribution, dependent on access to water for both manufacturing and transportation. Socially and economically, New York City relied on a vital and dynamic waterfront. The piers, slips, and basins of Lower Manhattan and Brooklyn, for example, were once the center of shipping and industry in New York's Upper Bay and the lower Hudson and East Rivers. But as both cities and industrial practices evolved in the mid-twentieth century, many formerly industrial urban waterfronts fell into disuse. The development of containerization in the 1950s precipitated the gradual movement of the shipping industry away from the Manhattan and Brooklyn waterfronts to the centralized New Jersey ports of Bayonne and Elizabeth. By the early 1960s, investment and expansion of other major American ports transformed the relationship between industrial shipping zones and their cities. By the 1970s, many formerly industrial urban waterfronts were abandoned and left to decay.

The postindustrial waterfront is no longer a working waterfront but a residential and recreational one, and although this re-envisioning has enriched cities, it has often reaffirmed the hardened coastline edge. Since the 1980s, city agencies, community groups, and private developers have rediscovered and reinterpreted these urban waterfront areas, transforming them into urban parks, recreational boardwalks, housing developments, and cultural institutions. In New York City and elsewhere, this transformation has occurred in parallel with emerging real estate markets. In the case of Battery Park City in Lower Manhattan, waterfront development not only renewed existing territory but reclaimed new land from the water by landfilling with the soil, rock, and sand from the adjacent World Trade Center excavation site. On a working waterfront, industrial shipping piers facilitated the exchange of goods and services—the passage of urban transactions across land and water. By contrast, the postindustrial waterfront conceals the daily and seasonal transformations of the coast, forming a crisp and aestheticized edge. Bold new forms and vistas often obscure the dynamic history of the waterfront, fixing the edge with a vertical seawall—an extruded line.

Today, the recognition of climate change, with its consequent sea level rise and ecological disruptions, encourages designers to look beyond a hard-edged waterfront and toward thoughtful adaptation to the subtle and dynamic processes that define urban shores. This demands representational techniques that trace the dynamic coastal processes—ecological and geological as well as urban and industrial—that shape the gradient territory between water and land. While engineering research and practice provides the essential representations—maps, equations, diagrams, digital models, physical models—through which the coast is understood, innovative representational tools derived from art and design practices are equally important for capturing the dynamism, flux, and indeterminism at work in the intersection of urban and natural systems along the coast. Where engineering practice seeks precision and fixity, defining and combining variables to yield replicable models, this design thinking emphasizes the role of abstraction in visualizing the coast in new ways. Holistic and integrated representations have the power to shift ingrained thinking about familiar sites. In the face of climate change, coastal resilience demands the cross-disciplinary invention of new visualization techniques.

This chapter illustrates an extraction of visual practices rooted in art, science, and engineering to create new methods of visualizing the coast. Twentieth-century art, geography, and engineering practices present excellent precedents, as do contemporary coastal resilience research and design projects. The first section, titled "Against the Waterfront: The Ecological Formless," outlines a framework for understanding and representing the confluence of human, natural, and industrial processes as they form the coastal landscape over long and short durations. The next section, "Measured Precision: Plan, Section, Topobathy," demonstrates how these multifaceted processes can be traced through the traditional tools of plan and section drawings as well as topobathy, which combines plan and sectional information to create a three-dimensional construction that bridges land and water. The third section, "Grids and Abstractions: Atlas, Matrix, Operation," delves into specific visual techniques that harness both the structure of gridded forms and the generative power of abstraction to illustrate the coast not as a hardened line or waterfront but rather as a site of diverse interactions between ecology, industry, and development. The final section, "Similitude, Sediment, and Scale: Digital and Physical Models," outlines the strengths and limitations of architectural and engineering models while advocating the use of hybrid models for coastal resilience work. These models draw from the abstraction of design models and the dynamism of hydraulic and wave simulations in order to illustrate the means by which dynamic processes can transform coastal regions over time. In sum, the chapter provides a toolbox of representational techniques that may be combined, extracted, or subverted for future inquiries into coastal flux.

Against the Waterfront: The Ecological Formless

If the hard-edged and constructed waterfront no longer serves as an adequate model for interpreting, visualizing, and reimagining the coast, new frameworks present an alternative. The coast demands a framework that prioritizes not finished products or stationary forms but generative processes and their capacity for making dynamic and reactive forms.

In the 1996 exhibition and subsequent book *Formless: A User's Guide*, art historians Yve-Alain Bois and Rosalind Krauss present a re-reading of modernist works through French intellectual Georges Bataille's theory of the *informe*. The formless, or *informe*, is not a "stable motif" but rather an "operation."[2] By definition unstable, the *informe* actively resists binary divisions of form and content. It embraces process, motion, energy, entropy, and material. The *informe* reveals what may have been repressed within a work of art or larger art historical narrative. To define and articulate the *informe* Bois and Krauss draw on the work of modernist and postmodernist artists with practices as diverse as Jackson Pollock, Richard Serra, Cindy Sherman, and Gordon Matta-Clark.

The *informe*, as defined by Bois and Krauss through Bataille, can be extended beyond artistic practice and into the landscape to define what might be called an ecological *informe*. Geological, hydrological, and atmospheric processes continually transform the earth's surface, informing and reforming the shape of land and water. Ecological systems then inhabit and interact with these topological surfaces. For example, strong winter winds move the "walking dunes" at Napeague Harbor in Montauk three and a half feet southeast each year, shifting vegetation patterns in their wake. On a larger scale, the Channeled Scablands of eastern Washington record the enormous glacial Missoula floods of the Pleistocene epoch. Similarly, the deep riverbed of the Hudson River and the underwater trench called the Hudson Canyon are the result of the termination and retreat of the Wisconsin Glaciation at the Verrazano Narrows more than 22,000 years ago. The ecological *informe* can also be identified in places where natural and human processes intersect, particularly in the era of the anthropocene. The anthropocentric processes of extraction, construction, and destruction have shaped the face of the earth as much as glaciers, rivers, and seismic forces.

The ecological *informe* combines an art historical narrative with an approach to environmental history that emphasizes the interaction between natural and human processes in forming the landscape. In his seminal 1996 essay "The Trouble with Wilderness: Or, Getting Back to the Wrong Nature," environmental historian William Cronon posits that "wilderness" is not so much a territory untouched by Western, and particularly American, civilization but rather a product of urban life. In tracing the myth of the wilderness from the nineteenth century through the modern environmental movement, Cronon argues that the long-held and pervasive American understanding of nature and wilderness as somehow distinct from and outside of contemporary society is ultimately damaging to the very environment that society aims to protect.[3] Cronon's argument for a new relationship to wilderness builds on earlier formulations of the interdependence of nature and industrialized society. In his 1964 book *The Machine in the Garden: Technology and the Pastoral Ideal in America*, literary scholar Leo Marx uses canonical literary works to trace the tension between a rural, pastoral ideal of America and the significance of industrialization in building the nation. For Marx, the reinvention of the pastoral ideal—the literary "garden"—offered a way to both comprehend and retreat from the rise of industrialization in the nineteenth century. Marx notes that this dynamic relationship between technology and pastoralism, though historically rooted, is not limited to a specific historical moment or even to literature; rather, "the contrast between the machine and the pastoral ideal dramatizes the great issue of our culture," forming an energetic and productive dialectic.[4] Likewise, Raymond Williams, in his essay "Ideas of

Nature," elaborates on the multiple and changing conceptions of nature, each defined in relation to "the idea of man in society," writing that "the idea of nature contains, though often unnoticed, an extraordinary amount of human history."[5]

The ecological *informe* thus provokes a twofold inquiry: first, an investigation into fluid and evolving natural and anthropogenic processes and the visual traces they imprint on the physical environment; and second, a reading of the fluid and evolving relationships between the natural world and the built environment as cultural narrative. The ecological *informe* straddles the physical and conceptual: The interpretation and illustration of physical processes reflects how intellectual and disciplinary boundaries are conceived. Thus defined, this ecological *informe* can be found in diverse visual practices ranging from art, landscape architecture, and geology, manifest in paper representations, and in direct engagement with the materials of the earth.

Entropy and the Ecological *Informe*

As both an environmental ethic and a visual language, the ecological *informe* provides a framework for representing landscapes defined by natural and industrial processes and for operating within those landscapes. Inspired by the industrial landscape and geological time scales, the work of sculptor and land artist Robert Smithson plays a critical role in defining the *informe* for Bois and Krauss and also serves as emblematic of the ecological *informe*. In the late 1960s and early 1970s, Smithson, who died at the age of 35 in a plane crash, produced a significant body of work that remains relevant to architects and landscape architects working in postindustrial urban and suburban contexts.

The denigration of the American landscape was, for Smithson, a source of creativity and invention. He revealed and reveled in the aesthetics of industrial sites and processes of decay. In his 1967 illustrated essay in *Artforum* titled "A Tour of the Monuments of Passaic, New Jersey," Smithson documents the "monuments" of this town—bridges, pumps, and drainpipes. Smithson was particularly taken with the machines and abutments along the Passaic River, describing the suburban industrial landscape as evident of a "prehistoric Machine Age" where buildings, especially new tract homes, "rise into ruin before they are built."[6] Smithson concludes the essay by describing entropy— a concept that fascinated him and inspired much of his work—through the example of a child playing in a sandbox, mixing black and white sand until it turns gray.[7] Smithson understood entropy as the "irreversible condition" of the "moving of forces toward a gradual equilibrium,"[8] but entropy, beyond its use as a description of a physical thermodynamic process, is also defined as the gradual decline into disorder. Passaic, with its decaying industrial past and postwar suburban neighborhoods, embodies that process of entropy.

The notion of the entropic in parallel with geological and fluvial processes is captured in Smithson's "flow" performances of 1969. In his October 1969 "flow," *Asphalt Rundown*, a dump truck pours asphalt down the slope of an abandoned dirt and gravel quarry outside Rome. In the 1993 film *Rundown*, Robert Fiori compiled footage of the mixing, pouring, and sliding of the asphalt down the coarse quarry cliff. As it flows downward, pulled by gravity, the thick material seeks the paths of least resistance, sliding into the fissures of the sandy ground. In *Rundown*, Smithson's wife, artist Nancy Holt, describes the asphalt "coagulating into the crevices of the earth" as "entropy made visible."[9] Smithson's 1969 drawing "Asphalt on Eroded Cliff" is as compelling as its large-scale dynamic earthwork counterpart. (See Figure 2.1.) With ink and chalk, Smithson captures the liquidity and energy of the asphalt

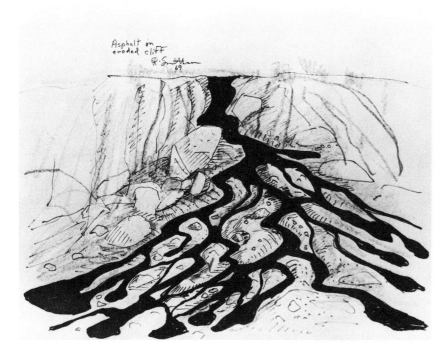

Figure 2.1: Robert Smithson, *Asphalt on Eroded Cliff*, 1969. Ink and colored chalk on paper, 18" × 24."

Courtesy James Cohan, New York/ Fred Jones Jr. Museum of Art, University of Oklahoma/© Holt-Smithson Foundation/Licensed by VAGA, New York NY

in motion, revealing how it shapes the quarry wall just as the asphalt is shaped by the wall itself. Smithson followed *Asphalt Rundown* with *Concrete Pour* in Chicago in November 1969 and *Glue Pour* in Vancouver in December 1969.

The submersion and then emergence of Smithson's well-known earthwork, *Spiral Jetty*, recalls the dynamic history of the waterfront piers of New York and other postindustrial cities. Since its construction in 1970, the jetty has charted and registered the entropic changes of the Great Salt Lake in Utah. For nearly 30 years the earthwork, constructed of piled rocks at the lakeshore, was submerged in the lake as the water rose. In 1999, when a drought caused the lake's water level to decrease, the *Spiral Jetty* emerged again, its rocks encrusted with salt. Time—whether moments or decades—drives entropy. Whereas fluctuating water cycles transform the jetty, cycles of urban development and economic and industrial activity similarly drive the decline, disappearance, and renewal of urban waterfront piers.

Industrial Decay and Designed Nature

Smithson's fascination with the interaction of natural and industrial processes was influenced by nineteenth-century landscape architect Frederick Law Olmsted, who, together with Calvert Vaux, designed New York's Central Park. Smithson acknowledges Olmsted as the first great earthwork artist in his 1973 essay "Frederick Law Olmsted and the Dialectical Landscape."[10] His reading of the "dialectic" landscape of Central Park—a giant swath of nature constructed in the middle of the continuous urban field of Manhattan—contains traces of its geological past while remaining an invention, a work of art. Smithson celebrates Olmsted's approach to landscape design not as static and formalist but instead as picturesque, defined by shifting relationships between objects and spectators and sensitive to time, change, and movement.

Indeed, Olmsted's work embodied the tension between the industrial and the natural in the context of the nineteenth century, yet its powerful programmatic elasticity has provided a powerful afterlife; it influenced Smithson in the 1970s and continues to influence contemporary landscape architectural theory and design. In her essay on Olmsted's career and legacy, historian Anne Whiston Spirn explains that Olmsted and his contemporaries in the mid- to late nineteenth century understood that nature could be constructed, and they constructed it wholeheartedly in New York and other American cities. But perhaps because Olmsted constructed nature so faithfully, later generations encountered his work as pure nature devoid of artifice, despite its designed and fabricated origins.[11] Spirn writes that "Olmsted represented a middle ground—which eroded in the twentieth century—between John Muir's nature as 'temple' and Gifford Pinchot's idea of nature as 'workshop.'"[12] One might argue that Olmsted saw nature, or his own designed natures, as a machine.

Smithson's fascination with Olmsted, Central Park, and the juxtaposition of the industrial with the natural world was realized physically in his sculpture "Floating Island to Travel Around Manhattan Island," sketched in 1970 but executed posthumously in 2005 by the Whitney Museum of American Art and the Minetta Brook Foundation. (See Figure 2.2.) A garden or park—or more precisely a conceptual extraction of Central Park—is planted within an industrial barge pulled by a tugboat. For 10 days in 2005, the "Floating Island" traveled around Manhattan during a Whitney Museum retrospective of Smithson's work. Landscape architect Diana Balmori designed the project from Smithson's sketch, with input from his wife, Nancy Holt. "Floating Island" was, for Smithson, a displacement or "non-site" of the similarly constructed landscape of Olmsted's Central Park.

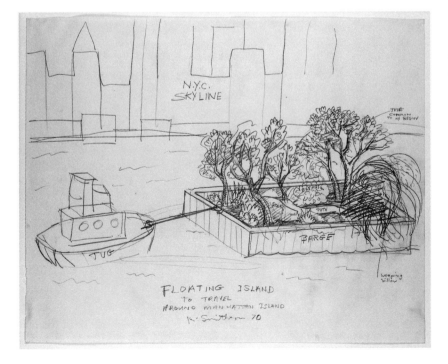

Figure 2.2:

Robert Smithson, *Study for Floating Island to Travel Around Manhattan Island*, 1970. Pencil on paper, 19" × 24."

Courtesy James Cohan, New York/ Private Collection/© Holt-Smithson Foundation/Licensed by VAGA, New York, NY

If in 1970 Smithson had intended the aesthetic juxtaposition of an industrial armature with a superimposed green landscape to disquiet or puzzle New Yorkers, when "Floating Island" circled the harbor in 2005 this juxtaposition had in fact become familiar. The Manhattan, Brooklyn, and Queens waterfronts, with their transformation from an industrial shipping interface to a residential and recreational edge, now precisely embody that aesthetic translocation. This transformation in meaning of Smithson's "Floating Island" between its conception in 1970 and its physical manifestation in 2005 signals the inevitable shift of cultural understandings of nature, landscape, and a reading of the waterfront over time. Art and design are critical to shaping these transformations.

Whereas "Floating Island" captures the aesthetics of nature and industry through precise design and construction, this relationship between industrial history and natural encroachment can be found in places untouched by artists and designers. At Shooters Island, an uninhabited 43-acre island off the northern shore of Staten Island along the Kill Van Kull, dynamic coastal processes are mixed and intermingled with those of industrial ruins. (See Figure 2.3.)

Figure 2.3: Aerial view of Shooters Island at the Kill Van Kull, north of Staten Island, New York.

© 2008 DigitalGlobe

A hunting preserve for wild geese during the Revolutionary War era, Shooters Island became a shipyard and oil refinery in the nineteenth century but was abandoned after World War I. Contemporary aerial imagery reveals slowly submerging docks and wooden ship hulls, now providing the foundation of a novel wetland habitat. The process of industrial decay transformed the island into a haven for bird species that returned to New York City in the 1970s after the Clean Water Act and a cleanup of the Hudson River estuary. The island, now maintained by the New York City Department of Parks and Recreation, is designated a "Forever Wild" bird sanctuary site. Shooters Island has thus returned to avian habitat—not through intentional restoration to a preindustrial ideal but through ecological appropriation occurring after a century of industrial use. The edges of Shooters Island are indeterminate; its western edge looks less like a coastline than a subsiding shipwreck. The dynamic industrial and postindustrial appropriations of the island remain present as traces, defining the form and texture of the land and providing a concise example of the ecological *informe*.

Mississippi Meander Belt

At a much larger scale, there is perhaps no better example of an ecological *informe* on the North American continent than the meander belt of the Mississippi River. Since the early twentieth century it has been confined to its path by engineered levees constructed for flood control and navigation. Before the era of flood control, the Mississippi changed course regularly, carving its way through the alluvial plain and depositing sediment during flood cycles to create rich agricultural soils and the intertidal marshlands of southern Louisiana. The river carries sediment toward the river delta and Gulf of Mexico, but as sediment is gradually deposited along the riverbed and banks, the river's current slows—until it shifts course to navigate a faster route to the sea. Historic channels of the Mississippi have left traces on the landscape, defining visible landscape patterns but also soil composition, informing agricultural use of the alluvial plain. The meander belt has captured the imagination of writers and artists alike, particularly John McPhee's chapter "Atchafalaya" in his 1990 book *The Control of Nature*.[13]

A stunning series of brightly colored geological maps produced in 1944 by geologist Harold Fisk for the U.S. Army Corps of Engineers (USACE) traces the overlap of nearly thirty historical channels of the river. (See Figure 2.4.) This sequence of tiled maps, extending from southern Illinois to

Figure 2.4: Harold N. Fisk, *Map of Ancient Courses of the Mississippi River Meander Belt (Cape Girardeau, MO–Donaldsonville, LA)*, Plate 22, Sheet 13, 1944.

United States Army Corps of Engineers

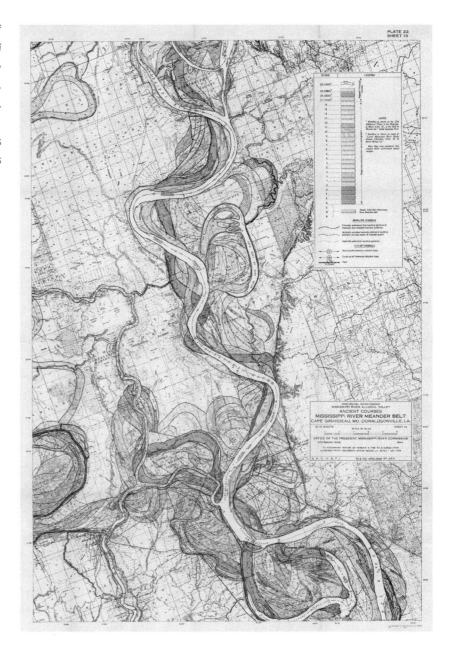

Louisiana, documents the sprawling geological meander belt of the Mississippi, many miles wide in some locations and strikingly narrow in others. Fisk and his team charted the four most recent courses—1944, 1880, 1820, and 1765—from historical surveys but also identified courses extending back over 4,000 years from aerial photographs, soil borings, and field surveys.[14] The land itself provides an archive of historical riverine paths. Fisk's "Map of Ancient Courses of the Mississippi River Meander Belt" captures this notion of the ecological *informe* not only because it documents a historical process of continual temporal and geological change and flux but also because it represents that process through an imaginative and convincing graphic language. The maps illustrate how the ecological *informe* spans both scientific and artistic languages, rendering visible and clear the earth's complex and dynamic geomorphology.

Visualizing the Ecological *Informe*

As the work of Smithson, Olmsted, and Fisk demonstrates, the ecological *informe* not only traces the interactions of human and natural processes but also illuminates new relationships through both careful observation and novel transformation. Contemporary artists and designers bring these processes into new light, providing inspiration for design practice while revealing the complexities of environmental degradation and climate change. Italian designer Bruno Munari's beautiful 1995 book *The Sea as a Craftsman* documents objects transformed by the ocean and washed ashore. (See Figure 2.5.) For Munari, the ocean is a designer capable of creating beautiful sculptures from human artifacts. But through the exquisite presentation and photography of his carefully collected objects, Munari reveals an unusual but productive collaboration between natural and human designers.[15] Similarly, Brooklyn artist Willis Elkins collects artifacts from Jamaica Bay and Newtown Creek, repurposing them as both art objects and tools. His "Jamaica Bay Pen Project" documents his collection of found and refurbished pens, as well as the drawings he produces from them. Elkins gives new life to these poignant sea-worn objects, which often bear the names of local businesses, some no longer extant, that surround the bay.[16]

Aesthetic practices that render visible the interconnected threads of the natural and constructed environment offer methods through which the

Figure 2.5: Atto Belloli Aldessi, in Bruno Munari, *The Sea as a Craftsman*, Corraini Edizioni, 1995.

Courtesy Atto Belloli Aldessi/© 1995 Bruno Munari/ Maurizio Corraini Srl.

ecological *informe* may be considered in the context of coastal resilience. The subsequent sections demonstrate a transformation of established representational techniques, with origins in architecture and engineering practice, that may be modified and manipulated to generate new ways of seeing and understanding the coast. Through the abstraction of drawings and models, designers can gain insight into dynamic processes, ultimately developing new forms of analysis and representation.

Measured Precision: Plan, Section, Topobathy

Three forms of representation—plans, sections, and topobathy—form the basis of coastal analysis by scientists, engineers, planners, designers, and government institutions. As technical data, these are the top-down tools of bureaucracy, rigid and prescribed. But through creative reinvention these representational tools can become instruments of exploration, invention, resistance, and change.

Architectural plans and sections provide idealized views that can be accurately measured and referenced for construction. Whereas a plan view captures a horizontal territory, a section provides detailed information through a vertical plane. Cut vertically through material constructions or geological topography, sections reveal the structural systems that comprise a building or landscape—the materials hidden behind cladding or below the surface of the ground. Plans and sections provide the basic representational tools for coastal engineering and flood mapping. When creating Federal Emergency Management Agency (FEMA) flood risk maps, the National Flood Insurance Program (NFIP) uses plan-based tools to model and represent the extent of flood inundation and sectional or transect-based tools to evaluate how waves will affect flood risk at regular intervals along a coastline. Coastal engineers also model the effects of waves on natural and constructed coastal structures by using transects or sectional profiles of the coast.

Three-dimensional topographic and bathymetric models combine horizontal and vertical information about a coastal site. Topobathymetry, or topobathy, is the merging of topographic data, or land elevation, and bathymetric data, or water depth. Topobathy models, which can be physical objects or digital files, visualize the continuous surface of the earth both above and below the water level. A topobathy is also essential to evaluating the effects of sea level rise and storm surge on coastal communities; this continuous surface is the vessel or container in which changes in water level can be assessed.

As static artifacts, plans, sections, and topobathy depict cities and regions at a single moment in the present, past, or future. But for coastal resilience, the examples that register change and dynamism can guide an understanding of the ecological *informe* in a specific location, revealing past and ongoing processes that might inform future change. Because plans—not only cartographic maps but also aerial photographs—operate at an urban or regional scale, they tend to reveal change over long periods of time. Sections cut along the coastline work at a closer scale, conveying the malleable exchange between land and water. When sequenced, sections show variation along the thickened edge, bringing clarity to a zone that might light look a line—or a waterfront—on a plan. In three dimensions, the topobathy creates continuity between topography and bathymetry, too often separated conceptually. Topobathic models register change beneath the water's surface, revealing the substrate necessary for building resilient coastal landscapes.

Plan

The bird's-eye view—represented in aerial photographs, plans, and maps—offers an expansive perspective on systems and spaces. The aerial view is a tool both for planning—for projecting new interventions and uses on the land—and for understanding the intersections of human and natural systems. Geographer Denis Cosgrove eloquently states that the aerial view affords two seemingly contradictory opportunities: on one hand, it provides a "canvas upon which the imagination can inscribe projects at a grandiose scale," and on the other hand, through its synthesis of vast landscape territory, it can "encourage a new sensitivity to the bonds that exist between humanity and the natural world."[17] Landscape architect James Corner puts it another way: "Paradoxically, the view from above induces both humility and a sense of omnipotent power."[18] This dual potential of the planar aerial view to cultivate understanding and inspire intervention is critical to coastal resilience. Plans offer insight into the relationships between natural and managed systems, allowing designers to determine where interventions might most successfully reduce flood hazards and improve ecological as well as public health.

The plan drawings, maps, and photographs described here illustrate the transformation of regions over time through the intersection and exchange of natural, urban, and industrial processes. When assembled they reveal site-specific urban narratives of ecological degradation, public health, agriculture, and development, all of which have profoundly affected the resilience, and vulnerability, of coastal communities. Through the superimposition of plans and plan-based data, researchers can uncover the ecological *informe* at work in specific sites.

Several nineteenth-century maps of New York City illustrate the critical, and long held, practice of discovering connections between the landscape and public health through cartographic investigation. The *Sanitary and Topographical Map of the City and Island of New York*, created by civil engineer Egbert L. Viele in 1865, documents the streams and wetlands that had been filled and capped by urban development, as well as constructed landfill at the edges of Manhattan Island. (See Figure 2.6.) Viele overlays the 1811 Commissioners' Plan for the rectangular street grid of Manhattan in order to provide a clear picture of the relationship between existing natural features and constructed development. Viele, an anticontagionist in the period before the germ theory of disease transmission was proven, was especially concerned with diseases he believed were caused by the miasma—noxious vapors or "bad air"—emerging from poorly drained lands, and his map identifies these low-lying marshy areas. But the real public health problems of an urbanizing New York in the late nineteenth century were waterborne and vector-borne diseases and the poor management of wastewater. By the end of the nineteenth century, the dumping of industrial waste and raw sewage into the

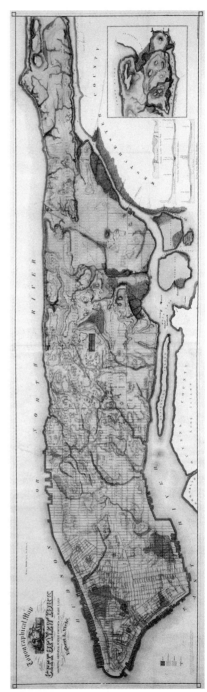

Figure 2.6: Egbert L. Viele, *Topographical map of the City of New York showing original watercourses and made land*, 1865.

New-York Historical Society, negative no. 74563

city's rivers and harbor had significantly damaged water quality and shellfish beds, causing serious typhoid outbreaks. A 1905 map and study produced by the New York Pollution Commission locates sewer outlets, shellfish beds, and tidal flow throughout the city's major waterways, indicating areas where the shellfish industry was most threatened by poor water quality. (See Figure 2.7.) This map, like that of Viele's earlier *Sanitary Map*, documents the tenuous relationship between urban and natural systems in a spatial and historical context.

Historical maps provide insight into ecological processes but also the larger-scale geological processes important to coastal resilience. In a different context, historic maps of China's Yangtze River delta at Shanghai document not only the transformation of the landscape through rapid urbanization but

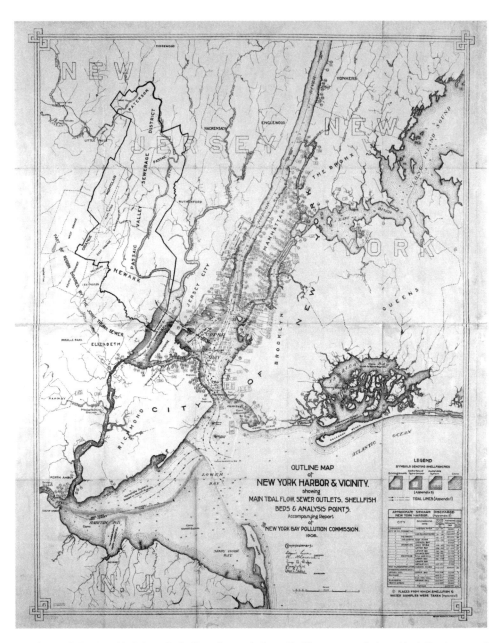

Figure 2.7: New York Bay Pollution Commission, *Outline map of New York Harbor and vicinity, showing main tidal flow, sewer outlets, shellfish beds and analysis points*, 1905.

Lionel Pincus and Princess Firyal Map Division, The New York Public Library

also the Yangtze River's gradual construction of the deltaic plain through sediment deposition and the parallel chenier ridges that extend toward the sea. The Yangtze River delta exemplifies the direct relationship between historical processes and contemporary coastal vulnerability; areas with rich histories of dynamic change are often the most susceptible to ecological damage and coastal hazards but also the most receptive to strategic design intervention.

A 1920 map of Shanghai and the Yangtze River delta by the British Whangpoo Conservancy Board illustrates the chenier ridges that characterize the relationship of the alluvial plain to emergent shallow-water wetlands. (See Figure 2.8.) The muddy river carries large amounts of sediment, which are deposited along the coast by the marine tides, creating earthen chenier ridges laden with shells parallel to the coastline. Vegetation gradually stabilizes these elevated ridges as they accrete outward, creating about 1 mile of new land every 70 years.[19] Like stepped terraces, the chenier ridges layer one after another, extending into the East China Sea. However, this low-lying topography of the deltaic plain is highly vulnerable to sea level rise and storm surge from tropical cyclones. Recent efforts to protect land from flooding through the construction of extensive seawalls has damaged and in some areas destroyed the wetlands formed by the Yangtze's deposition of fluvial sediment. The Whangpoo Conservancy map not only provides insight into contemporary vulnerability of the region but also offers inspiration for alternative methods of protecting the city beyond seawalls.

As individual artifacts, historic maps capture a moment in time, but when combined, compared and overlaid, they chronicle the interdependent human and natural processes that together transform coastal regions. Catherine Seavitt's work for *Structures of Coastal Resilience* includes an analysis of six National Oceanographic and Atmospheric Association (NOAA) charts dating from 1879 that register significant topographic and bathymetric changes in and around Jamaica Bay, New York. (See Figure 2.9.) The tracings of these changes create a new plan of multiple histories, identifying shifts in the shape of the land and its shoreline over time. (See Figure 2.10.) By the early twentieth century, channel deepening and dredging of Jamaica Bay not only allowed commercial ships to enter the bay but also yielded fill material for the construction of the Floyd Bennett airfield. The 1933 stone jetty constructed by the USACE at the western end of the Rockaway Peninsula encouraged sand accretion and the creation of new land for what is now known as the Breezy Point neighborhood. Deep borrow pits still remain in the formerly shallow

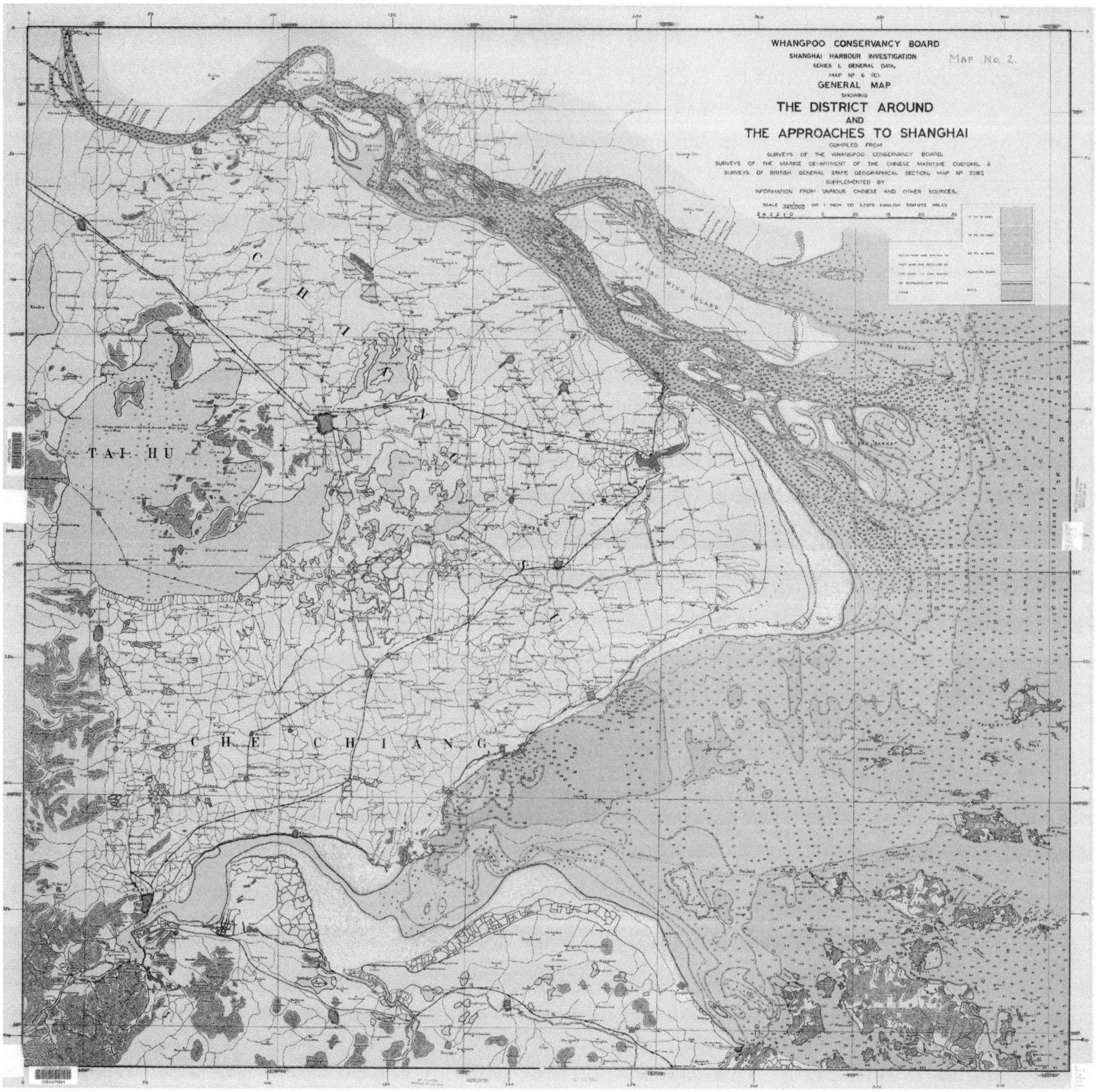

Figure 2.8: Whangpoo Conservancy Board, *General map showing the district around and the approaches to Shanghai*, 1920

Earth Sciences and Map Library, University of California, Berkeley

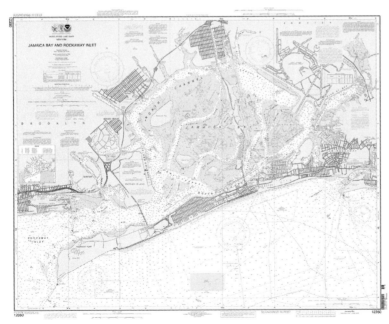

Figure 2.9: United States Coast and Geodetic Survey, *Jamaica Bay and Rockaway Inlet*, 1903 (top).

National Oceanic and Atmospheric Administration, *Jamaica Bay and Rockaway Inlet*, 2011 (bottom).

National Oceanic and Atmospheric Administration

Figure 2.10: Shifts in coastal edges at Jamaica Bay between 1879 and 1929, produced from historical NOAA nautical charts.

Catherine Seavitt/City College of New York, *Structures of Coastal Resilience*, 2015

Grassy Bay area, where sand was extracted and displaced to create the airfield of the John F. Kennedy International Airport, a former marsh. The NOAA maps illustrate encroaching development in the areas surrounding the bay, documenting the urban transformation of this outer extent of New York City.

These dramatic shifts in topography and bathymetry have ecological ramifications, many of which have been captured with aerial photography. In contrast to navigational charts, which clearly articulate distinctions between land and water, aerial photographs capture the murky exchange between sea and city. Taken from 1924 through 2014, a series of aerial photographs reveals both the short- and long-term transformations of Jamaica Bay's marsh islands. (See Figure 2.11.) Each image captures the bay at a unique moment in the tidal cycle, but across time a general trend emerges: The islands shrink as marshland disappears, a result of multiple factors occurring in

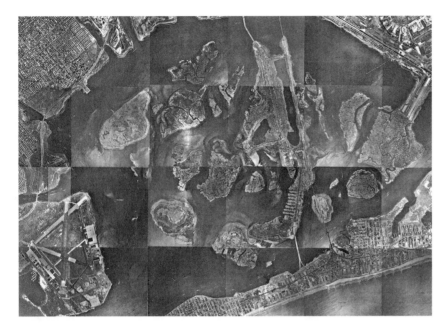

Figure 2.11: Aerial photographs of Jamaica Bay from 1974 (top) and 2013 (bottom) reveal salt marsh loss within the bay.

New York State Department of Environmental Conservation and © 2013 DigitalGlobe

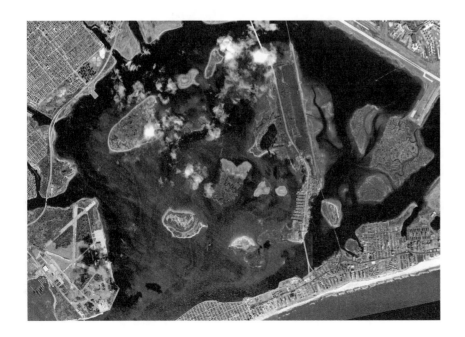

combination, including poor water quality, urban development, and sea level rise. The aerial bird's-eye or plan view offers an image of the coast from above; it provides a totalizing picture that captures many aspects of a coastal region—tidal zones, vegetated areas, roads, bridges, neighborhoods. (See Figures 2.12 and 2.13.) However, aerial photographs, like maps, obscure other important information about ecological health: water depth, soil composition, the height of trees, the density of marsh grasses. Aerial photographs, and plans more broadly, are just one piece of a larger portrait of ecological and geological change in a region.

Although historic maps and aerial photographs contribute to our understanding of the present, researchers must go further to identify, map, and overlay contemporary data that drives resilience in a given community. Like Viele in the nineteenth century, designers can create new plans that synthesize the types of vulnerability that contribute to ecological health, flood risk, and social resilience. Seavitt's work for Jamaica Bay includes a series of

Figure 2.12: Aerial view of Little Egg Marsh, Big Egg Marsh, and Broad Channel at Jamaica Bay, 2014.

© Vertigo Aerial Imagery for *Structures of Coastal Resilience*, Jamaica Bay

Figure 2.13: Aerial view of the West Pond breach at the Jamaica Bay Wildlife Refuge, 2014.

© Vertigo Aerial Imagery for *Structures of Coastal Resilience,* Jamaica Bay

plans that map the environmental, social, and infrastructural vulnerability of Jamaica Bay, identifying zones where design might yield multifaceted solutions, bringing economic development, social resilience, or environmental rehabilitation to areas prone to flooding and its consequences. A plan of salt marsh loss between 1879 and 2011 visualizes the extensive loss of wetlands due to development, shipping channels, poor water quality, and rising sea levels. (See Figure 2.14.) Areas of particularly high environmental sensitivity— zones that house endangered or threatened animal and plant species—are also mapped. Mapping social and demographic data is often particularly revealing. A plan of population and risk factors combines population density with characteristics that often make residents more vulnerable to catastrophic events such as floods. Organized by census tracts, these metrics identify households with older adults living alone, families living under the poverty line, residents with limited English, and single mothers with children. Finally, a plan of

Figure 2.14: Salt marsh loss in Jamaica Bay, with data drawn from NOAA historic nautical charts from 1879, 1948, and 2011.

Catherine Seavitt/City College of New York, *Structures of Coastal Resilience*, 2015

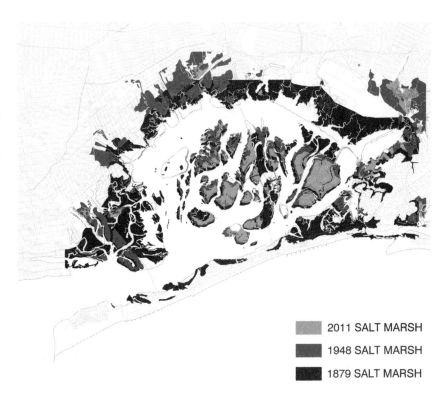

2011 SALT MARSH

1948 SALT MARSH

1879 SALT MARSH

infrastructure (major roads, evacuation centers, power supply stations, water facilities, and police and fire stations), with the hurricane evacuation zones superimposed, reveals the extensive public infrastructure systems vulnerable to flooding during a storm event, as well as the neighborhoods that will be isolated when vulnerable infrastructural systems are compromised.

Section

Whereas plans capture large horizontal territories, sections denote vertical information along a single transect. Sections are not limited to architectural representation; they are an essential tool for measuring and modeling coastal engineering systems. Sections clarify the geomorphology, material composition, and construction techniques within a coastal formation. At a larger scale, sections can illustrate the full gradient coastal zone from high ground inland to intertidal or submerged ground offshore, showing typical or specific conditions. And when sequenced, a series of sections can demonstrate the transformation of conditions along a coastline, showing variation and continuity in a coastal region. Section drawings are a critical tool for both engineers and architects, as they provide measured, precise documentation in the vertical plane. But they can also be highly visual representations of urban experience, providing imaginative imagery not only of how the city might be constructed but also how it might look and feel at a human scale. This dual capacity makes the section an important tool for understanding the changing nature of the coast and its capacity for future change.

Coastal engineers evaluate cross-shore slices or transects of the coastal zone to study processes such as wave action, erosion, sediment transport, and littoral drift, as well as the effects of these processes on coastal structures such as dunes, jetties, groins, and seawalls. Beach profiles—measured line drawings showing vertical dimensions from offshore bathymetry through beach zones and shore protection structures—are a key tool for evaluating environmental change between seasons and over extended time periods. Researchers might take field measurements several times a year at specific beach profile markers along a shoreline. To create flood risk maps, FEMA flood modelers use the Wave Height Analysis for Flood Insurance Studies (WHAFIS) software for selected coastal transects. Topography, vegetation, and structures along a transect line are entered into the program, which then calculates how waves will affect flooding along that two-dimensional sectional profile.

Coastal engineers also interpret sectional information in three dimensions, conducting physical experiments in wave tanks and flumes, long, narrow rectangular pools that can simulate thickened sections of the coast. In a wave flume, a wave-maker or oscillator device at one end of the long water tank generates waves, which are observed by scientists and measured by digital sensors as they propagate toward the other end. The USACE Engineer Research and Development Center (ERDC) in Vicksburg, Mississippi houses 1.5- and 3-meter-wide glass-walled wave flumes, both 63 meters long, to test sediment transport and the effects of waves and surge on coastal structures. The largest indoor wave research facility in the country belongs to Oregon State University; the O. H. Hinsdale Wave Research Laboratory's large flume measures 104 meters long by 3.7 meters wide and 4.6 meters deep. During a celebrated experiment at Hinsdale, organized by the USACE in 1991 and titled SUPERTANK, an enormous sand beach was constructed in the tank to study storm-induced erosion. At 203 meters long and 20 meters wide, the outdoor saltwater wave flume at the Ohmsett facility in Leonardo, New Jersey is designed specifically for oil spill research. Operated and maintained by the U.S. Department of the Interior's Bureau of Safety and Environmental Enforcement, Ohmsett's wave flume includes a moveable wave-damping

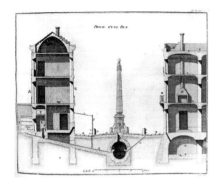

Figure 2.15: Pierre Patte, "Profil d'une rue" ("Section through a street"), from *Mémoire sur les objets les plus importants de l'architecture*, Chez Rozet, Paris, 1769.

artificial beach.[20] Wave flumes make transects physical, but they must still be understood as sectional information. Like section drawings, they capture one isolated moment of the coast and project information gained from that transect to more complex sites.

The section drawing also has a long history of documenting the role of infrastructure and technology in cities not only at the coast but beneath the streets and behind the facades of buildings. A historic tradition of visionary urban street sections originated in the eighteenth century. In 1769, French architect Pierre Patte published a drawing titled "Profile d'une rue," or "Street Section," in his book on building technology and construction. (See Figure 2.15.) The drawing cuts through a typical Parisian street to reveal a design for an underground sewer system connecting to the street above and buildings on either side. By the nineteenth century, street sections became a standard form of representing urban technology. Etchings from the era of Baron Haussmann's 1850s and 1860s era in Paris illustrate sewers under boulevards as well as the life inside typical Haussmannian apartment buildings. In the early twentieth century, French architect and urban planner Eugène Hénard visualized layers of underground automobile and streetcar traffic in street sections that detailed the many elements of urban technology, from coal delivery to drainage to street lighting. These sections explain the city as an integrated mechanism, dependent on technology that traverses both public and private space.

Following from this lineage, section drawings can visualize the integration of coastal resilience into the existing systems of urban life, above and below the streets and sidewalks. Street section drawings of lower Manhattan produced by Architecture Research Office (ARO) and DLANDstudio can be understood within this lineage of urban sections; they reveal landscape strategies such as pervious streets, planting beds, and soil composition as urban technology. Created for the Museum of Modern Art (MoMA) exhibition *Rising Currents*, a series of illustrated sections shows how porous streets and green spaces in Lower Manhattan could become buffers and storage zones for storm surge and stormwater, improving drainage and reducing potential flood damage. (See Figure 2.16.) ARO's sectional drawings of downtown streets and plazas—West Street, Water Street, Broadway, Hanover Square, and Coenties Slip—demonstrate how city streets can be reconfigured to absorb and slow water through interventions both above and below the ground. The drawings suggest condensing underground utility conduits into waterproof vaults beneath the sidewalk, ensuring infrastructural protection during storms while allowing the streets to become absorbent systems with

Figure 2.16: Transformation of subgrade infrastructure with accessible waterproof utility vaults under sidewalks and permeable streets to absorb excess stormwater.

Architecture Research Office and DLANDstudio, MoMA *Rising Currents*, 2010

layers of soil and gravel. Importantly, these sections also reveal the infrastructure needed to make green spaces possible; a section of Coenties Slip shows the substantial retaining walls needed to sustain a submerged park.

At a larger scale, sections that traverse land and water can illustrate how the edge condition, or gradient waterfront zone, might be sensitively transformed to engage with the existing structure of the city. For the master plan in the research book *On the Water: Palisade Bay*, the precursor to their *Rising Currents* proposal, ARO divided the Lower Manhattan coastline into zones, each with a distinctive formal composition of buildings, parkland, piers, and elevated highways. (See Figure 2.17.) At Battery Park City, an existing seawall becomes a higher but gently sloping vegetated earthen berm that provides tidal wetland habitat. The proposed section for Battery Park extends the park farther into the water but replaces the existing seawall with tidal pools and

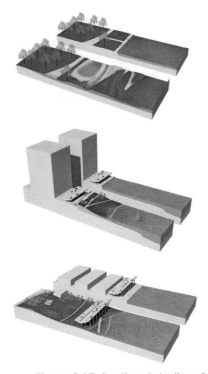

Figure 2.17: Sectional studies of existing and proposed conditions along the Lower Manhattan waterfront. The widening of landscape at Battery Park diverts high tides and storm surge (top). A berm parallel to the elevated Franklin D. Roosevelt East River Drive protects inland development and creates a new elevated pedestrian boardwalk (middle). Open space behind the highway can filter stormwater runoff and absorb surge floodwaters (bottom).

Guy Nordenson, Catherine Seavitt, and Adam Yarinsky, *On the Water: Palisade Bay*, 2010

strategically placed berms. At the elevated Franklin D. Roosevelt East River Drive, the existing vertical seawall is transformed into a new berm and park that gently slopes into the water. And just south of the Brooklyn Bridge, a new open green space filters stormwater runoff underneath the elevated highway.

At an urban or regional scale, sequences of sections can show variation in coastal conditions, illustrating the diversity of a coast. For their work with *Structures of Coastal Resilience*, Anuradha Mathur and Dilip da Cunha of the University of Pennsylvania produced a set of sectional sketches highlighting the existing variation along the Virginia coast. Mathur and da Cunha identify four coastal conditions extending from Norfolk: the James River to the west; the Great Dismal Swamp Canal, the Intracoastal Waterway, and the Elizabeth River to the south; the coastline of Virginia Beach to the east; and the eastern shore of Chesapeake Bay to the north. They then chronicle a series of typological sections along each of these coastlines with line drawings, showing the topographic shifts, structures, and vegetation that characterize each region. From this analysis they design new sectional sequences for each area, transforming existing conditions with resilient features. These sectional drawings illustrate both ecological and programmatic gradients that extend both inland and offshore. (See Figure 2.18.) One section for the "Fall Line" along the James River documents the transition from salt marsh to lowland forest to upland forest along a gradual slope. A section for the Great Dismal Swamp Canal, part of the Intracoastal Waterway, proposes a series of stepped berms that provide habitat for plant and animal species suited to a particular elevation. A third example, for the eastern shore of Chesapeake Bay, presents land-building troughs in a channel off Tangier Island. Mathur and da Cunha's sections demonstrate the many ways that land can transition into water—the multitude of conditions that characterize a resilient coast.

Topobathy

Whereas sections show the transition from land to water along a single vertical plane, topobathy models combine sectional and planar information to provide measured documentation of the continuous ground surface at the scale of a community or region; they provide vertical elevation data over a wide horizontal extent. These models are instrumental in conveying the continuous relationship between land and water and the land beneath the water. With roots in technical cartographic practice, topobathy models are numerical models containing precise elevation data at a large scale. But precisely because they represent the earth's surface so literally and in a form not com-

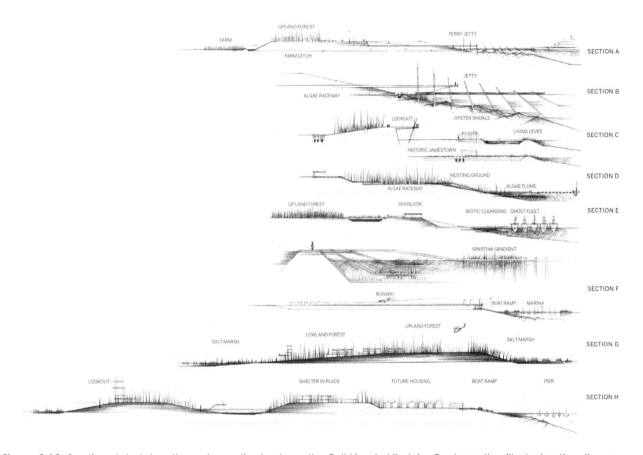

Figure 2.18: Sectional sketches through gradients along the Fall Line in Virginia. Each section illustrates the diverse transitions from water to land, and from high ground to low ground, in the region.

Anuradha Mathur and Dilip da Cunha/University of Pennsylvania, *Structures of Coastal Resilience*, 2015

monly visualized, topobathy models ultimately provide a new abstraction of the coast that is particularly generative for coastal resilience.

Because it encompasses elevation data both on land and under water, topobathy requires a combination of cartographic methods. As a technical tool, topobathy models are often digital files created and manipulated in geographic information system (GIS) software. Typically, topographic and bathymetric data are stored as digital elevation model (DEM) rasters—pixelated image files that contain an elevation value for each square pixel. (See Figure 2.19.) The resolution of elevation data varies, depending on the source and production method of the data files. New topographic and bathymetric data

Figure 2.19: Digital elevation model (DEM) showing topography and bathymetry of Jamaica Bay. Darker zones indicate deeper terrain, and lighter zones indicate higher elevations.

Catherine Seavitt/City College of New York, *Structures of Coastal Resilience*, 2015

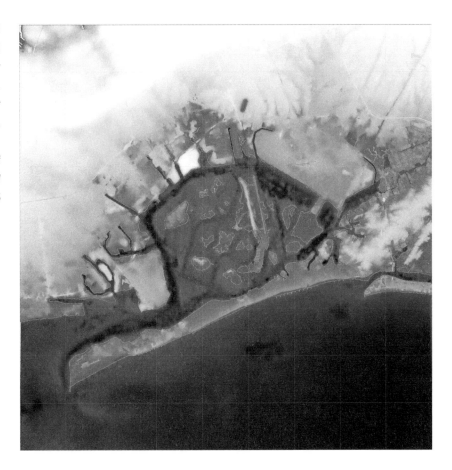

are often created by LiDAR, a remote sensing technology that measures distance with laser pulses.[21] LiDAR sensors, which are usually airborne via airplane or helicopter, collect precise, high-resolution elevation data. Historical methods of collecting topographic information include traditional surveying techniques and the use of interpreted aerial imagery. Bathymetry is also collected through acoustic sonar; a device mounted on the side of a boat releases soundings, which then hit the ocean floor and bounce back. The time needed for the sounding to travel can be translated into a depth measurement.

Making a topobathy from distinct digital models of land and water involves constructing the coastal zone in three dimensions, reconciling distinct representations to bring clarity and precision to the coast. Because they traverse land and water, topobathy models are often compiled from distinct sets of source data. Topographic and bathymetric data sets are produced separately by different agencies; the U.S. Geological Survey (USGS) distributes topographic data, and NOAA generates bathymetric data. These data sets must be combined to make an accurate digital topobathy file for a specific region, and this process requires adjusting datums to find an accurate shared reference datum. Yet, with sea level rise, coastal zones are shifting. Continuous digital topobathy models are essential to modeling the effects of sea level rise and storm surge on a coastal community. Design interventions can also be modeled and merged into a topobathy model, and the flood mitigation effect of these insertions can be tested with sea level and surge scenarios.

The significance of topobathy for coastal resilience is greater than the technical capabilities for modeling and mapping that it makes possible, however. By articulating the gradient between the ocean floor and land above sea level, topobathy allows a reinterpretation, and reimagination, of the continuity of the coast. Typically relegated to technical practices, topobathy models have only recently made their way into the fields of architecture and landscape architecture. Coastal resilience work has made topobathy a critical tool, not only for understanding coastal geology but also for conveying the three-dimensional shape of coastal regions to a general audience. Three-dimensional visualizations of topobathy models can be powerful communicative tools. In 2010, Guy Nordenson and Catherine Seavitt created two large physical topobathy models of the Mississippi River delta and the New York Upper Bay with precisely this intention. (See Figures 2.20 and 2.21.) These models were part of an exhibition, created in partnership with the Louisiana State University Coastal Sustainability Studio, for the United States Pavilion at the 2010 Venice Biennale. These physical models depicted the continuous topography and bathymetry of terrain, independent of the water volume. The water itself, constructed with layers of transparent acrylic, was extracted as a volume and suspended above the gray topobathy below, thus causing the coastline to disappear and revealing these two river estuaries as vessels. Positioned side by side, the models emphasized the distinct topographic conditions of the Mississippi River delta and the New York Upper Bay at the Hudson River estuary. The Upper Bay, created by glacial movement, exhibits deep fjord geology—the bay is flanked by a geological palisade to the west and the emergent bedrock schist of Manhattan to the north. By contrast, the Mississippi delta is flat and vast, comprised of soft sedimentary soils carved by the once-meandering, but now confined, deep Mississippi River channel. Positioned side by side, the two models reveal clearly the fallibility of considering the coast as a line as well as the unique geomorphology of different coastal regions.

Figure 2.20: Mississippi River delta topographic/bathymetric model installed at *Workshopping: An American Model of Architectural Practice*, United States Pavilion, 12th International Architecture Exhibition, Venice Biennale, 2010.

Catherine Seavitt and Guy Nordenson/© Resnicow Schroeder Associates Inc., photo Daniele Resini

Figure 2.21: New York/New Jersey Upper Bay topographic/bathymetric model installed at *Workshopping: An American Model of Architectural Practice*, United States Pavilion, 12th International Architecture Exhibition, Venice Biennale, 2010.

Catherine Seavitt and Guy Nordenson/© Resnicow Schroeder Associates Inc., photo Daniele Resini

The traditional representational tools of plan, section, and topobathy may thus be reimagined as dynamic documents, revealing the flux of a coastal ecological *informe*. Historically conceived as static tools, these examples describe new and novel forms of systems thinking brought to representational media, allowing a new approach to visualizing, designing, and engaging coastal resiliency as an active process.

Grids and Abstractions: Atlas, Matrix, Operation

Another familiar form of visual representation, the geometric grid, imposes mathematical, measurable structures on places, images, and concepts that are subject to other, often less tangible structures or systems. Grids operate as representational tools that also come to define the physical geographic territory they represent, as evidenced by many examples including the geographic coordinate system of longitude and latitude, the 1785 Jeffersonian grid that defines American land ownership, and the street grid of New York City as defined by the Commissioners' Plan of 1811. Grids provide a means for understanding, and developing, the world. But grids can also have the opposite effect, underscoring the power of visual representation to influence perception. Through the superimposition of a structure without hierarchy, grids can flatten preconceived notions of a region or concept, allowing new hierarchies and ideas to reveal themselves. The grid is a critical emblem of modernist art. In her seminal 1979 essay "Grids," Rosalind Krauss argues that grids in modernist art function "to declare the modernity of art . . . flattened, geometricized, ordered. . . . [The grid] is what art looks like when it turns its back on nature."[22] Yet Krauss also argues that grids allow multiplicities of meaning that encompass science and spirituality, mathematics and myth. The grid allows precision, but it also creates distance; this dual capacity makes it ripe for interpreting the coast and the ecological *informe*.

Grids are crucial to the generative process of defamiliarization, a process with origins in artistic practice. The homogeneity of presentation can paradoxically draw attention to the subtle differences between images. This effect is evident in the mid-century photographs of industrial architecture by German photographers Hilla and Bernd Becher. The Bechers consistently photographed their subjects—coke ovens, water towers, grain elevators, and oil refineries—under specific lighting conditions, framing each structure on center to ensure consistency across images. (See Figure 2.22.) The clinical presentation of their work, with images of similar buildings arrayed in

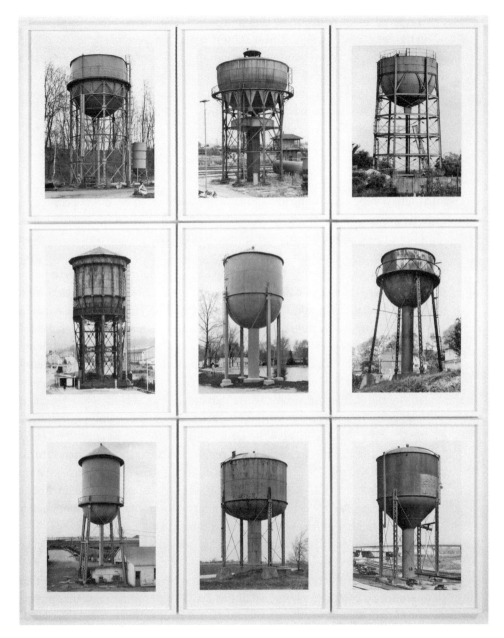

Figure 2.22: Bernd and Hilla Becher, *Water Towers*, 1980. Nine gelatin silver prints, 155.6 cm × 125.1 cm.

Courtesy Solomon R. Guggenheim Museum Foundation/© 1980 Bernd and Hilla Becher/Licensed by SCALA/Art Resource, New York

structured grids, allows the differences between images to emerge.[23] The methods of analysis and representation presented here use grids to derive logical interpretations while also generating abstractions, which might in turn produce new ideas for coastal resilience. These methods demonstrate confidence in ordered, taxonomic systems. Yet the power of the grid lies not in its ability to compute a single, unequivocal solution but rather in its generative capacity to catalyze new design ideas.

Atlas

An atlas compiles a sequence of maps into a compilation or collection and in so doing catalogs them through a gridded hierarchy. A classic road atlas of the United States, for example, presents each state on a new page, clipping and cutting up the country into a nonetheless logical book that is useful to a driver. A world atlas might similarly divide the globe into pages defined by continents or countries or longitudinal coordinates. An atlas imposes its own organizational and graphic logic onto a geographic region in order to make sense of it. But this cartographic rigor also allows room for artistic defamiliarization, as any new graphic system enables new readings of existing territory.

Through the imposition of a regular grid on an aerial image, designers can transform a coastal region with a new spatial hierarchy. Guy Nordenson and Catherine Seavitt created an "edge atlas" for the New York and New Jersey Upper Harbor for *On the Water: Palisade Bay* and again for the mouth of the Yangtze River at the East China Sea for the *Yangtze River Delta Project*. As an exhaustive record of a geographic area, the edge atlas affords designers a unique opportunity to identify regions that require remediation or design intervention. Because satellite images of vast geographic territories can obfuscate coastal details, dividing a large coastal territory into smaller segments yields a more precise understanding of a site.

Nordenson and Seavitt adapted the representational technique of the edge atlas from a 1989 book of maps, *An Atlas of Venice*. In 1982, the city commissioned a series of aerial photographs, which were carefully stitched together to produce a complete photomap of the city. This photomap was then subdivided into 186 squares along a regular grid. Cartographers traced the photomap of each square area to create an analogous line drawing, encoded with geometric and topographic details: Buildings are outlined in red and trees in green; line weights and types signify canals, bridges, staircases, and railways; and dashed lines demarcate surfaces such as pavement, lawns, and playing fields. Each spread of the bound volume juxtaposes a line drawing on the left-hand page with a square photomap segment on the right-hand page. The line drawing reveals ambiguous conditions, and the photomap offers colorful detail and depth.[24]

Inspired by *An Atlas of Venice*, an edge atlas requires three steps: First, a regular grid is overlaid on a large aerial photograph, dividing the site into square segments; second, each of these segments is examined in two and three dimensions using aerial photography to produce a diagrammatic overlay that captures the extent of floodplains, land use, and the shoreline's structural composition; third, unadulterated satellite images and diagrammatic data are assembled into book spreads. The consistent spread layout places a diagram on one side and an aerial image on the other for a direct side-by-side comparison.

For the edge atlas of *On the Water: Palisade Bay*, a ten-square by ten-square grid was imposed on a satellite image of the New York/New Jersey Upper Harbor. (See Figure 2.23.) The grid squares that touch both land and water are assigned a number and an individual spread; in total, forty-eight blocks circumscribe the bay and the islands within it. Although some of the grid squares depict neighborhoods, industrial sites, and parks only partially adjacent to the water, the floodplain data demonstrate that inland low-lying regions may be as susceptible to inundation as land directly adjacent to the water. Other squares reveal only a sliver of land beside an expanse of water. While the outline of the harbor takes on a recognizable shape, the abstract division of the site into uniform squares offers a fresh perspective on familiar territory.

As an index of existing conditions, the edge atlas is a tool for anticipating and preparing for flooding due to sea level rise and storm surge from tropical cyclones. The edge atlas technique captures the diversity of shoreline conditions along an urbanized coast. A single coastline might include an industrial shipping yard, a residential development, and parkland, and each is adjacent to a varying shoreline type—seawalls, breakwaters, revetments, or wetlands. Dynamic natural processes of erosion, drift, and accretion, combined with human appropriations and interventions, yield an environment of remarkable ecological and programmatic heterogeneity. For the *Yangtze River Delta Project* edge atlas, Seavitt developed a thorough system of diagramming probabilistic floodplains and edge conditions for each square grid cell. A color gradient illustrates the likelihood of flooding: orange tones represent the 250-year flood zone (0.4 % annual chance of annual flooding) and

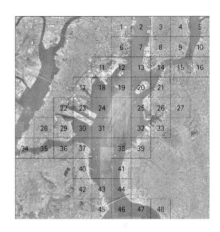

Figure 2.23: The edge atlas divides the New York/New Jersey Upper Harbor into an even grid of squares, capturing an array of coastal conditions.

Guy Nordenson, Catherine Seavitt, and Adam Yarinsky, *On the Water: Palisade Bay*, 2010

deep red tones designate the 1,000-year flood zone (0.1% chance of annual flooding). A line along the boundary between water and land indicates one of seven different edge conditions by color—polder water, polder drained, upland vegetation, agricultural, wetland, channel, or hardscape—and an overlaid hatched line represents the varying height of seawalls, designed to protect against 50-, 100-, 250-, or 1,000-year floods. (See Figure 2.24.) These diagrammatic notations are overlaid on the corresponding satellite images and placed next to the original image, revealing the most vulnerable regions and the conditions that create these vulnerabilities.

Matrix

Whereas an edge atlas imposes a geographic gridded structure on a spatial region to show an existing territory in new light, a matrix uses a conceptual or visual grid structure to generate new combinations of forms and ideas. Paul Lewis of Princeton University used a speculative matrix for his *Structures of Coastal Resilience* proposal for Atlantic City, New Jersey as both a representational tool and a productive design strategy for exploring the hybridization of natural, structural, and urban features to generate new formal structures of coastal resilience. (See Color Plate 2.) With the use of an autonomous, systematic framework to develop novel forms, Lewis advances an established architectural practice in the formalist tradition, using geometric manipulation to generate formal possibilities.

The speculative matrix draws on methodological traditions in mathematics, biology, and ecology, but it also participates in a recent trend to

Figure 2.24: A grid square from the *Yangtze River Delta Project* edge atlas charts existing coastal edge conditions and storm surge vulnerability.

Catherine Seavitt and Guy Nordenson, *Yangtze River Delta Project*, 2013

incorporate matrix thinking into landscape architecture. A mathematical matrix organizes and tracks data that depend on the combination of multiple parameters. Since the early twentieth century, geneticists have used the Punnett square, perhaps the most recognizable scientific matrix, to graphically represent the probabilistic results of breeding experiments. Each hybrid square contains both a string of letters that correspond to the genes that the hybrid might contain and an illustration of how the organism might appear as a result. By contrast, the ecological matrix takes on a different meaning as the connective tissue of an environment. The matrix is still a structuring framework, but it is also the context in which all systems exist and processes occur. In recent decades, matrices have become a common tool for the landscape architect because, as Linda Pollak explains in her essay "Matrix Landscape: Construction of Identity in the Large Park," landscape architects have engaged and collapsed the multiple definitions of matrix, effectively adapting the graphic matrix from mathematics and genetics to communicate and reconcile the dynamic processes at play within the ecological matrix.[25] This interest in matrices stems in part from the work of the landscape architect Ian McHarg, who expanded the domain of landscape architecture to a regional scale. In a graphic, tabular analog to his vibrantly colored and layered maps in his groundbreaking 1969 book *Design with Nature*, McHarg uses matrices to examine the compatibility of various land uses—urban, suburban, agricultural, and recreational. Prospective land uses are listed along both axes of the matrix so that each may be considered in relation to all others. McHarg determined where each land use fell on a four-degree scale, ranging from incompatibility to full compatibility, and graphically denoted this compatibility with a colored dot at each intersection of the matrix.[26] McHarg sought a systematic, science-based approach to reach objective solutions to design problems, but Pollak argues that contemporary landscape architects have begun to combine the graphic matrix with other representational techniques, such as collage, to move beyond the limited solutions suggested by McHarg's diagrams and toward a plurality of original design possibilities.

Building on this trend in landscape architecture, Lewis's speculative matrix also draws on recent USACE initiatives. Historically the USACE has relied on rigid, highly engineered structural systems to guide and contain riverine channels and protect coastal areas. Partly in response to the devastating failure of these methods during Hurricanes Katrina and Sandy, USACE has recently introduced natural and nature-based features (NNBFs)—elements such as dunes, wetlands, living shorelines, and barrier islands—to improve coastal storm risk management. Together with more traditional nonstructural and structural interventions, these NNBFs describe the "full array of measures" that USACE can implement to support coastal resilience and storm risk management. To couple NNBFs with structural solutions, USACE has developed a series of "combined profiles," which aggregate various measures through adjacency.[27] For example, a submerged breakwater, a seawall, and wetland vegetation are placed in a linear sequence.

Lewis and his team turned to their speculative matrix as a tool to systematically integrate the NNBF and structural coastal measures of the USACE with well-known urban typologies. Whereas traditional protection structures segregate the city from the coast, the pairing of structural measures with nature-based features yields a range of intriguing new conditions at the coastal edge. Additionally, NNBF measures can be combined with urban features to transform water into an urban amenity and an integral part of daily life.

The vertical axis of the speculative matrix lists two sets of landscape features, nature-based and urban. Nature-based features include forests, wetlands, reefs, beaches and dunes, and barrier islands, whereas urban elements include the block, street, yard or sidewalk, building, and boardwalk. The horizontal axis arrays flood protection features, both nature-based and structural. Elements from the vertical and horizontal axes are crossed to create a suite of sixty hybrids, each of which couples a structural or nature-based flood protection measure with an urban or nature-based landscape feature.

One column illustrates the hybrids produced by the combination of wetlands and urban elements, strategically introducing water and vegetation into urban areas. Combining a neighborhood block with a wetland yields an urban condition where the streets and alleyways become canals. Combining wetlands and streets results in a similar scenario, yet the roads become estuarine habitat rather than open water. The wetland/yard illustration demonstrates the possibility of a typical expanse of turfgrass surrounding a house transforming into marsh grasses and a sidewalk into an elevated boardwalk. The hybrid feature at the intersection of buildings and wetlands illustrates the adaptation of a home's foundation to a wetland condition, either as a sunken bunker or a raised platform. Finally, the boardwalk/wetland image shows a parklike space with a raised wooden walkway, meandering into coastal waters. Each of these scenarios demonstrates the potential of an urban system absorbing and retaining higher water levels.

By contrast, the row that hybridizes the city block with nature-based features and structural elements contains a series of scenarios that are provocative but less practicable. Crossing a block with a breakwater yields a bold geometric pattern but not a functional alternative to existing structures. Likewise, a hybrid of the city block and a beach/dune produces a whimsical illustration of mounds of sand, each bound within a grid of streets; yet as a design solution, the condition has limited applicability. Other conditions generated by the matrix, including the combination of blocks with bulkheads and revetments, lead to the existing back-bay condition of the Chelsea Heights neighborhood, the site of the team's final design proposal.

For hybrids that demonstrate particular promise as resilient design solutions, Lewis introduces a third element. For example, a breakwater is added to the block and wetland hybrid to illustrate how the wetland/block typology can transition to open water. (See Color Plate 2.) The matrix provides an efficient means for developing hybrid coastal conditions, allowing the designers to focus less on the initial generation of ideas and more on refining how each test case might be formally and functionally realized.

A matrix not only visualizes how geological, biotic, and human systems interact but offers a framework for understanding the indeterminacy of the natural environment at coastal sites. The dynamism of the coastal landscape poses a particular representational challenge; tidal cycles set the water in perpetual motion, and the land shifts in response. A multitude of plant and animal species live, feed, and breed along the water's edge, adding to the complexity of the coast. Finally, people are drawn to coasts, adding industrial, residential, and recreational development. The landscape embodies the matrix; it is a substrate within which systems cross, interact, and hybridize. Yet just as the graphic matrix fails to produce a single determinate scenario that captures all inputs, the coastal landscape offers a profusion of uncertain and indeterminate outcomes.

Operation

When combined with techniques of collage, grids provide organizing structure to coastal analysis while allowing unique forms of representation inspired by the places they interpret. The analytic work of Anuradha Mathur and Dilip da Cunha of the University of Pennsylvania for their site of the Tidewater region of Virginia for *Structures of Coastal Resilience* demonstrates how structured collages can create visual representations that reconstruct regions to shed light on the social, economic, and ecological characteristics of the coast. The collage techniques addressed here draw from art practice, but they use specific maps and scientific graphics as source material. Abstraction becomes a tool not for undermining this information but rather for generating new meaning and new representations of the coast, which can ultimately yield design concepts.

Whereas Nordenson and Seavitt rely on the adjacency of diagrams and aerial imagery in their edge atlas to highlight programmatic and material variations along the water's edge, Mathur and da Cunha reorder the crenellated coast, turning the water into a guiding straight line down the center of a series of vertical collages. The result draws out variations in land use and edge conditions along the coast that would be overlooked in an aerial photograph. Mathur and da Cunha cut and array satellite images to generate new composite images for interfaces extending outward from Norfolk in four directions: the Fall Line escarpment along the James River to the west, the Intracoastal Waterway from the Great Dismal Swamp to Craney Island to the south, the beachfront along Virginia Beach to the east, and the eastern shore of Chesapeake Bay to the north. (See Color Plate 3.) To create these long vertical collages, Mathur and da Cunha array aerial images of the landscape into strips perpendicular to the shore of the ocean, river, or bay. They then realign these strips to transform the coast into a straight edge. The collage of the Intracoastal Waterway, for example, reveals the transformation of the coastline from a swampy vegetated edge in the south toward an industrialized coast in the north, where the canal intersects the Elizabeth River and eventually the Chesapeake Bay. Moreover, a comparison between the four collages accentuates similarities and differences between coastal conditions within the Tidewater region, revealing latent opportunities for design.

In another analytic collage technique for *Structures of Coastal Resilience*, Mathur and da Cunha aggregate photographs, maps, and drawings to suggest the interactions between the natural, cultural, infrastructural, and historical narratives of sites in Tidewater Virginia. Their "Beach Front Research Plot" outlines the process of coastal erosion and dune formation alongside images of Hurricane Irene and meteorological visualizations of past hurricanes that affected the area. (See Figure 2.25.) These depictions of disaster and destruction are juxtaposed with a map of the Virginia Beach boardwalk, license plates with coastal imagery, tourist spending data, and advertisements for beach vacations. A line of maps at the bottom of the composition demonstrates the shrinking of the beach over time. The collection of these images into a single

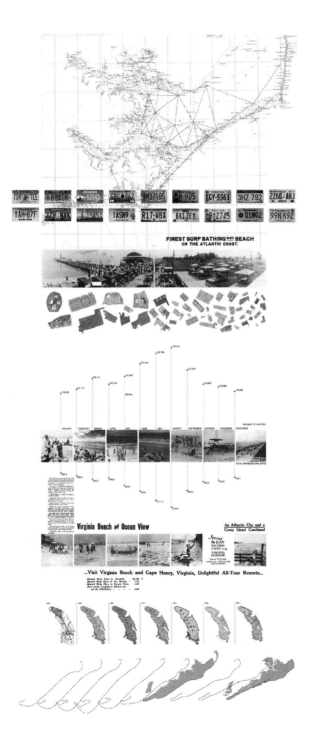

Figure 2.25: The "Beach Front Research Plot" compiles historical data, photographs, and maps to create a visual portrait of the region's vulnerability to hurricanes and storm surge.

Anuradha Mathur and Dilip da Cunha/University of Pennsylvania, *Structures of Coastal Resilience*, 2015

composite facilitates a deeper understanding of the relationships between the ecological consequences of a disaster and the cultural and economic character of a region. The "Eastern Shore Research Plot" similarly illustrates the relationship of agriculture, aquaculture, pollution, and landforms along the Chesapeake Bay through an assemblage of historical charts, photographs, and newspaper clippings. (See Figure 2.26.)

By assembling images and information, Mathur and da Cunha explore the tensions inherent in combining natural and human-made systems in design and construction. A series of gridded collages titled "Operational Gradients" explores the process and effects of building and managing areas of constructed high ground in the Norfolk area. Through the combination of graphic impressions and detailed drawings, these collages illustrate the tensions between four site-specific relationships: dredged channels and high ground, upland and tidal flooding, species adaptation and water quality management, and short-term tactics and long-term strategies for managing flood risk. (See Color Plate 4.) "Dredging Channel/Building Ground" highlights dredging channels for navigation and building high ground for flood risk management. Photographs demonstrate the diversity of soil types in the region—submerged land below the sea, sand in the tidal zone, and soil piled high above sea level—and images of soil remediation tactics explain the treatment needed at each elevation along the gradient. For example, dredged material from Craney Island must be sorted, treated, remediated, and mixed with other material to eventually facilitate urban agriculture, forestry, or development. The collage also includes channel sections, a regional plan, and requisite plant materials for each type of soil remediation.

As this work demonstrates, collages can be analytic but also operational; the act of collecting and arranging disparate images can produce new meaning through adjacencies. It also showcases how design thinking can draw from both art and science. Art historian Aby Warburg's *Mnemosyne Atlas* is an early example of the juxtaposition of images as a generative tool. Left unfinished at the time of his death in 1929, the *Atlas* illustrates how images that held great symbolic importance or emotional power in antiquity reveal themselves in the art and cosmology of later periods. (See Figure 2.27.) Warburg's piece consisted of sixty-three wooden panels draped in black cloth and covered with thematically related photographs, clippings from newspapers and magazines, and pages of text. The panels could be rearranged to yield an ever-growing set of nested and interconnected histories. Contemporary scholars continue to construct new interpretations of

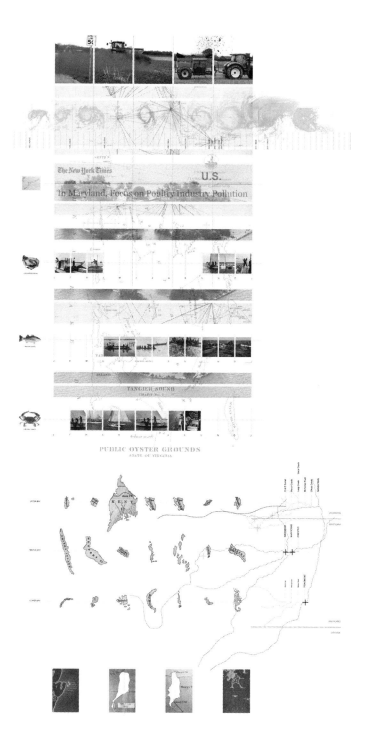

Figure 2.26: The "Eastern Shore Research Plot" of the Chesapeake Bay compiles the competing interests, industries, and ecologies in a region characterized by rivers, sounds, and islands as well as poultry farming and agricultural fields.

Anuradha Mathur and Dilip da Cunha/University of Pennsylvania, *Structures of Coastal Resilience*, 2015

Figure 2.27: Aby Warburg,
Mnemosyne Atlas, Panel 8,
1924–1929.

Courtesy The Warburg Institute

Warburg's panels, proving the inexhaustible potential of image multiples.[28] Mathur and da Cunha's collages similarly produce meaning as both process and product; they assist the designers in developing coastal resilience projects but also encourage viewers to generate new interpretations of dynamic sites.

Mathur and da Cunha's site collages for Tidewater Virginia develop from a long tradition of experimental representation in landscape architecture that combines artistic visualization with scientific information to suggest, with fixed imagery, the ever-changing and evolving nature of landscape. The pictorial collage, with its perspectival basis, has a particularly long history: Even before landscape architecture was recognized as discipline, eighteenth-century gardeners experimented with spatial compositions and material palettes through collage. In the 1790s, gardener Humphry Repton layered watercolor illustrations of existing and proposed conditions so that the drawing of the existing condition could be folded back to reveal his vision for the future landscape.[29] In the twentieth century, landscape architects appropriated methods of collage and montage from visual and filmic arts. In contrast to photorealistic rendering, collage facilitates the creative process through fragmentation and ambiguity. The collages of landscape architect Yves Brunier, dating from the 1980s, offer one example of this approach; the layering of media in his work is not only visual but also tactile, offering new sensory experiences.

Mathur and da Cunha's recent coastal resilience collages follow from a defining moment in the 1990s when collage became a new and important analytic device in landscape architecture. Importantly, this rise of collage paralleled the increasing disciplinary interest in industrial processes and postindustrial landscapes. This moment is perhaps best captured by James Corner's 1996 book *Taking Measures across the American Landscape*, which combines maps, photographs, and drawings to describe the reciprocal relationship between natural geological formations and human land use. In "The Survey Landscape Accrued," Corner layers segments of aerial photographs, topographic maps, line drawings of natural features, and an oblique view of farmland onto the Public Land Survey System grid, demonstrating how a regularizing system can counterintuitively yield variation in the landscape. (See Figure 2.28.) Corner's collages are both projective and productive: Various parts of the collage provoke, instigate, and suggest new possibilities in both spatial and temporal planes. Whereas Corner's representations emphasize the influence of human-made systems and structures on the American landscape, Mathur and da Cunha choose to accentuate the power of natural processes on developed and industrialized landscapes with their collages. For their 1997 book *Mississippi Floods*, Mathur and da Cunha traveled the Mississippi River valley, studying the ecology and cultural landscape of the region to provide a unique interpretation of this great American river that emerges from design thinking and artistic imagination. They collected traditional forms of design representation—maps, diagrams, and sections—but also photographs, media reports, and paintings. Screen printing offered the design team a unique way to combine these elements; the production method mimicked both the natural additive process of sedimentation and the layering of cultural conceptions over time. In contrast with the crisp and geometric lines of Corner's collages, the *Mississippi Floods* screen prints are expressive and impressionistic. As an analytic tool, they convey the indeterminacy and intensity of the processes involved in the creation of the Mississippi River valley. (See Figure 2.29.)

Atlases, matrices, and collages operate as productive analytic tools when applied to the dynamic environment of the coast. As organizational techniques borrowed from geographic, biological, and artistic practices, these strategies not only analyze the indeterminate coast but, through a process of defamiliarization, allow designers to project new resilient design ideas into temporal systems.

Similitude, Sediment, and Scale: Digital and Physical Models

The twenty-first century is an age of models, particularly digital ones—economic models, financial models, weather models, climate models, building information modeling (BIM) models, biological models—but also physical, three-dimensional ones. Models represent reality, but to do so within physical or computational limits, they must abstract that reality. They depend on abstraction to provide clarity. To produce precise quantitative or qualitative information, models must simplify and reduce the complexity of inputs. In his book on climate modeling, *A Vast Machine,* Paul Edwards offers a simple rebuttal: "Without models, there are no data."[30] All knowledge of climate is gained through models—from daily weather forecasts to the most elaborate climate simulations. In order to understand the world, it must first be abstracted.

Models are quintessential tools of both design and scientific practice, and despite disciplinary differences, they are used in diverse fields to explain not

Figure 2.28: James Corner, "The Survey Landscape Accrued," in *Taking Measures across the American Landscape,* Princeton Architectural Press, 1996.

James Corner

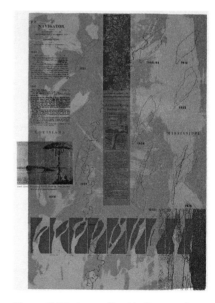

Figure 2.29: Anuradha Mathur and Dilip da Cunha, "Lateral Moves," from *Mississippi Floods: Designing a Shifting Landscape,* Yale University Press, 2001.

Anuradha Mathur and Dilip da Cunha

only how things look but also how they work, or could work. Architects and landscape architects use scalar models to convey the spatial characteristics of a project—how geometric planes create interior voids and exterior forms, and how those forms relate to the surrounding built environment. Models can also illustrate the materiality of a building or landscape and the effect of that material on its spatial, functional, and emotive character. The model can be a presentation piece, a precision object exhibiting the final state of a project, but it can also be a valuable working tool during the design process; it might be transformed and reconfigured to allow the designer to both think and experiment in three dimensions.

Whereas the traditional architectural model is static, physical hydraulic engineering models incorporate kinetic elements—flowing water, current-producing paddles, suspended sediment loads, and erodible topography—to illustrate the interaction between the landscape and the water that flows through it. Engineering models strive for both geometric and kinematic similarity such that precise, quantitative measurements from test trials might be used to design and locate dams, levees, and other water control structures. A physical model can reveal unexpected interactions that might be missed in a digital model, particularly because of the reduction of input variables in the digital modeling platform. Although architectural models illustrate the character and qualities of a place through material and texture and form, they often lack the dynamic components necessary for representing physical change in coastal contexts. Engineering models capture this dynamism. But to do so they isolate variables, testing specific processes apart from the natural, social, and cultural contexts in which they operate. In other words, both types of models fall short in considering the diverse components of coastal resilience.

Recent coastal resiliency projects developed by designers to confront regional flooding and storm conditions—challenges historically solved by engineers—have used a type of hybrid model that couples the architectural model's ability to experiment with form and communicate concepts with the dynamic range of the engineering model. Through descriptive observation, photographic and video recording, and digital sensing, these hybrid models capture the changes in a system over time. They reveal the physical processes that govern coastal systems and offer designers a path toward developing practicable, resilient interventions. As a working tool but also a critique of disciplinary isolation in coastal resilience, hybrid models capture the ecological *informe*, in both spirit and practice.

Material as Metaphor

Physical models made by architects and landscape architects for coastal projects abstract complex contexts to represent design interventions that may be subtly integrated into an urban fabric. Models reduce and simplify physical information to convey design concepts as well as the material and textural language of these interventions. The tactile qualities of physical models can capture the attention of a wide public audience without a background in design. But in the context of coastal resilience, static architectural models rely on metaphor to convey the dynamic processes and temporal changes essential to coastal resilience.

The 2010 MoMA workshop and exhibition *Rising Currents: Projects for New York's Waterfront* offers an example of architects and landscape architects conveying coastal resilience concepts to a broad public through physical models built with various materials. The MoMA has a rich history of encouraging designers to engage with the city and address urban problems and then presenting their ideas to a public audience through physical objects, especially architectural models. The MoMA Department of Architecture and Design helped catalyze public housing reform with *Housing Exhibition of the City of New York* (1934) and later tackled redevelopment with *The New City: Architecture and Urban Renewal* (1967).[31] *Rising Currents* followed in this activist tradition, explaining to the public that New York City will be increasingly susceptible to catastrophic flooding due to sea level rise and more frequent extreme weather events, both precipitated by climate change. Physical models, the physical and conceptual centerpiece of the exhibition gallery, met this challenge through specific material, construction, and presentation techniques, each representing a site physically not only through visual reproduction but also with abstraction. Five design teams created models through careful choice of material, scale, color, and projection to illustrate visions of a new New York City, one that would potentially be both greener and more ecologically vibrant.

Through the contrast between muted opaque white acrylic buildings and bright translucent green materials at the ground plane, ARO and DLANDstudio's scale model of *New Urban Ground* emphasized the capacity of the urban streetscape to mitigate flooding through a mesh of concrete and plants resilient to saltwater intrusion along the perimeter of the island. This green connective tissue would absorb and drain rainwater while providing a buffer from storm surge during severe weather events.

The effervescent green of this model of Lower Manhattan contrasted with the rough-edged model of SCAPE's *Oyster-tecture* reef, constructed from milled plywood, wooden dowels, and layered webs of crocheted yarn. (See Figure 2.30.) Simple silhouettes of people, boats, birds, and marine species punctuate this matted field, suggesting an underwater reef supporting a rich texture of human and marine activities. Emergent marsh grass filled the zones between the layers of crocheted webs, illustrating the biological dynamism of the underwater context.

Four sectional models by Matthew Baird Architects for *Working Waterline* reveal the industrial program of the site in Bayonne, New Jersey with existing piers and open water mediated by jack-shaped recycled glass breakwaters. The murky hues and distressed finishes evoke the industrial character of an oil tank farm transformed into a glass recycling facility, converting discarded glass into the building blocks of a coastal reef designed to attenuate wave energy. The models collapse the hard boundary between land and water, expressing the physical, economic, and ecological connections across the gradient section.

For *New Aqueous City*, nARCHITECTS created two large models that hung on the gallery wall. One represented the full project site on either side of the Verrazano Narrows in layered acrylic; a second model at a larger scale articulated the fabric of an amphibious urban neighborhood with mono-

Figure 2.30: Detail of a hand-knitted model of the *Oyster-tecture* reef at the Bay Ridge flats offshore Sunset Park, Brooklyn.

Kate Orff/SCAPE Landscape Architecture PLLC, MoMA *Rising Currents*, 2010/© 2010 MoMA, photo Ian Allen

chrome white cardstock in shallow relief. (See Figure 2.31.) Whereas the former served as a three-dimensional master plan, with the specific design elements (island, piers, barriers, housing, and ferry routes) rendered in bright, distinct colors, the latter demonstrated the extension of new aqueous housing and water treatment wetlands from the existing city grid and urban fabric. The contrast between these two models—one brightly colored and smooth and the other monochromatic and subtly textured—suggested that the proposed design would transform the structure of the city while respecting existing models of urban form.

Digital projection can transform a static model into a dynamic representation tool. LTL Architects' model for *Water Proving Ground* shows the rise and fall of the sea around the landscape of Liberty State Park in Jersey City. The white low-relief milled model of the proposed piers and buildings registered the topographic variation across the park landscape, and a sequence of images projected onto the model from above illustrated how the site could be activated during high and low tides and distinct flood conditions, accommodating changes in sea level over time. (See Figure 2.32.) Together, the

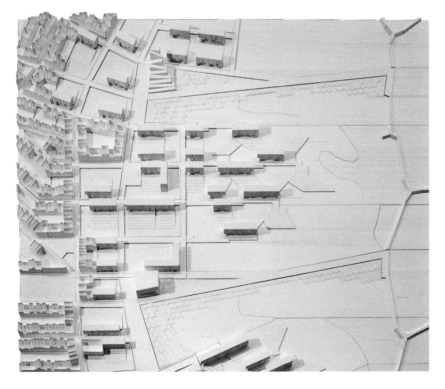

Figure 2.31: Axonometric model of *New Aqueous City* with a detail at Staten Island, showing island networks, inflatable flood barriers, piers, and inverted housing.

Mimi Hoang and Eric Bunge / nARCHITECTS, MoMA *Rising Currents*, 2010/© 2010 MoMA, photo Frank Oudeman

Figure 2.32: *Water Proving Ground* model for the Jersey City shoreline, with projected images of fluctuating tides and surge flooding.

Paul Lewis, Marc Tsurumaki, David J. Lewis/LTL Architects, MoMA *Rising Currents*, 2010/© 2010 MoMA, photo Ian Allen

architectural and landscape models created for *Rising Currents* presented a new visual image of a resilient city.

Similitude and Precision

If architectural models present design arguments through material, scale, color, and projection, engineering models also test and present arguments about how processes work. Qualitative and quantitative measurements serve as tools of inquiry and forms of evidence. Hydraulic engineers model the behavior of water in space and time. They use large-scale hydraulic prototypes to demonstrate how water acts and reacts at specific junctures and small-scale physical models to study flow within riverine and estuarine systems. Coastal engineers also model coastal processes including wave action, erosion, dune formation, and wave attenuation. Engineering models depend on the abstraction of contexts and inputs in much the same way as architectural models. They test phenomena in conditions that may approximate, but do not replicate, full-scale riverbeds or coastal regions. Models depend on the isolation of variables; abstraction is critical to yielding informative results. To produce useful information, models must reduce, simplify, or ignore the full spectrum of interconnected riverine or coastal processes.

The challenge is how to acknowledge and learn from this abstraction in a field that seeks similitude and precision as the ultimate goal. Engineering models are evaluated according to similitude, the ability of a model system to accurately describe the behavior of a comparable system outside of the lab. Geometric similitude requires equivalency in the ratio of the corresponding linear dimensions, length, area, and volume. Architectural models typically achieve this metric. Engineering models used to investigate processes of hydraulic flow and sediment deposition must also demonstrate kinematic similitude, the proportionality of elements including time, velocity, acceleration, and discharge. Dynamic similitude combines geometric and kinematic similitudes; a model with dynamic similitude can be used to predict the outcomes of interventions in coastal and riverine planning. In his 1687 *Principia* Sir Isaac Newton outlined the "Principle of Similitude": In order to demonstrate similar behavior, two systems must be not only geometrically proportional but proportional in time, velocity, mass, and acting forces.[32] This simple observation, made more than three centuries ago, provides the theoretical foundation of present-day hydraulic engineering models.

This emphasis on similitude drove centuries of research into hydraulic models. For nearly two millennia, engineers have sought means to create greater geometric and kinematic similitude. The earliest recorded hydraulic model was built by Sextus Julius Frontinus, the Roman senator responsible for the city's water supply under Emperor Trajan around 60 CE. He created a scale model of the aqueduct system to determine where sand was being deposited and thus develop a strategy to combat this deposition.[33] Throughout the nineteenth century, steady technological advancement enabled the design of increasingly precise models and the collection of ever more exact data. British naval engineer William Froude sought to design ships that would move through the water safely and efficiently. Because of the complexity and inaccuracy of the calculations involved in mathematically predicting the drag of a ship, Froude constructed large models of typical hulls and floated them through long narrow tanks to formulate laws of water resistance.[34] Through his research he developed a formula, the Froude number, which relates the behavior of similar objects at different scales. French engineer Louis Fargue, a contemporary of Froude, pioneered the use of physical modeling to study river hydraulics in the late nineteenth century.[35] During the next half-century others refined Fargue's methods. Working in England, Osborne Reynolds built a scale model of the Mersey estuary to study eddies caused by tidal currents. His experiments depended on understanding the ratio between inertial and viscous forces, later known as the Reynolds number, which could be used to predict the dynamic similitude between modeled and real-world conditions.[36]

Mississippi Basin Model

The history of large-scale hydraulic models in the United States follows from the Mississippi River flood of 1927, which precipitated the large-scale Mississippi Basin Model (MBM). The construction, use, and eventual abandonment of this hydraulic model offers insight into the ambition for physical models in twentieth-century engineering practice but also the challenges they present. The aspiration toward similitude can overshadow a greater, more subtle understanding of the complexity of large-scale landscape systems.

In the early twentieth century, German and American engineers developed an intellectual exchange and built increasingly advanced experimental hydraulics laboratories. Upon his return from a 1913 trip to Germany, American engineer John Freeman lobbied for a national hydraulics lab in the United States. It was not until the devastating Great Flood of 1927, however, that Freeman prevailed. The year 1929 marked the opening of the first federal hydraulics research facilities: the National Hydraulic Laboratory in Washington, D.C. and the Waterways Experiment Station (WES) in Vicksburg, Mississippi.[37]

After experimenting with a series of outdoor hydraulic models in the 1930s, the USACE began construction in 1943 on the largest hydraulic model in the world to date.[38] The project represented 1,250,000 square miles of the Mississippi River basin as a 1:2,000 scale model, built on 800 acres on a rural site 40 miles from the WES. (See Figures 2.33 and 2.34.) The scale model included 15,000 miles of the Mississippi River and its major tributaries, the Tennessee, Arkansas, and Missouri Rivers. However, it stopped at Baton Rouge; the lower Mississippi from Baton Rouge to the Gulf was never modeled.[39] The MBM, which took more than 20 years to complete, accurately simulated the conditions of the 1952 and 1973 floods and was used to determine the placement of locks, dams, and flood control structures.[40] The model

Figure 2.33: Section of the abandoned Mississippi River Basin Model, Clinton, Mississippi, 2014.

Photo Danae Alessi

Figure 2.34: Detail of abandoned Mississippi River Basin Model, Clinton, Mississippi, 2014.

Photo Danae Alessi

underscored the immensity, complexity, and highly interconnected nature of the river system. It also demonstrated the potential fallibility of relying on the brute force of levees and flood walls to control flooding at isolated points.[41]

In the postwar period, USACE constructed a series of other hydraulic models for estuaries as varied as the Savannah harbor, the Delaware River, Charleston Harbor, Narragansett Bay, and Chesapeake Bay. These hydraulic models tested fluid dynamics but also enabled experiments regarding water composition and salinity. In 1956 the USACE built a scale model of the San

Francisco Bay and Sacramento River delta to test the Reber Plan, a proposal to fill in much of the bay and construct a deepened shipping channel for defense. Accurate salinity in the model's water made it possible to test saltwater intrusion alongside sediment movement and flood control.[42] The New York Harbor Model, in operation from 1957 to 1965, reproduced tides, currents, shoaling, and storm surge and also tested how radioactive waste might travel through the estuary.[43]

Digital Hydraulic Models

The decline in the use of physical riverine models followed from the development of digital tools understood to be even more precise than their physical counterparts. By the 1970s, computer programs that could predict and analyze riverine conditions provided an alternative to physical models, which were expensive to maintain and often took weeks to run and analyze. Outdoor locations also made it difficult to ensure consistent baseline conditions. In 1971 the Mississippi Basin Model received funds to conduct a 2-year computer application study to compare the results of the river model with data produced by new software developed by the USACE's newest branch, the Hydrologic Engineering Center (HEC). As a result of this turn toward digital models, the MBM was used only sporadically throughout the 1980s and abandoned by 1990.[44]

Although large-scale river models fell out of favor with the advent of new hydrodynamic digital models, the USACE continued to use smaller hydraulic models integrated with computer systems to improve their precision. Since 1994 the USACE has used micromodels, also called hydraulic sediment response (HSR) models, which are small-scale, tabletop replicas designed to explore the effects of control structures on sediment transport and river flow.[45] HSR models consist of a polyurethane foam insert, accurately formed using georeferenced aerial photography, placed within a table outfitted with electronic control valves, pumps, and flow meters that are run and monitored by a computer system. Lasers gather detailed bathymetry and velocity information that can be compared with acoustic Doppler current profiles (ADCPs) from the actual river. High-definition cameras observe the model, visualizing flow and deposition.[46]

Although structures, topography, flow rates, and velocities can be accurately scaled down, the physical properties of water do not scale, leading to distortions in properties such as viscosity and surface tension.[47] Many geometric and kinematic variables are also difficult to reproduce with accuracy at such a small scale.[48] But despite these limitations, physical models remain an irreplaceable tool. They can demonstrate physical processes that are not yet well enough understood to be incorporated into computer simulations.[49] Although digital models effectively simulate closed systems, they are ill-suited to complex, open systems. In the past decade there has been a renewed interest in physical models to capture the dynamic variables of coastal resilience projects. Pairing physical modeling techniques with technology refined in HSR models, small-scale physical models are reemerging in major research facilities from Brazil to Louisiana.[50] Used in combination with numerical or computation models, they aid in achieving a rigorous and holistic understanding of coastal and river hydrologic systems.

In 2013, Louisiana State University and The Water Institute of the Gulf announced plans to construct a Water Campus in Baton Rouge. The facility will include a $16 million small-scale physical model (SSPM) of the Mississippi River at a 1:12,000 horizontal and a 1:500 vertical scale, capturing 190 miles from Donaldsonville, Louisiana to the Gulf of Mexico. More than 200 5-foot by 10-foot high-density foam panels, milled using a computerized numerical control (CNC) router, comprise the base. The model will be located indoors so that it will be possible to carefully control experimental conditions and maintain the 90-foot by 120-foot model. Once again, a Mississippi River model will be one of the largest hydraulic models in the world.[51] The model will use precisely selected and calibrated synthetic sand to simulate the sediment transport characteristics of the Mississippi River. Experimental results from this new Mississippi River model will be used in tandem with numerical models and computer simulations to help guide the planning and design of river diversions. The model also runs much faster than previous study models; an entire year of river time runs in only 20 minutes. Decades of sediment diversion can be simulated in a matter of days. Field data related to sediment deposition and dredging volumes have been used to verify the results of a test model. Hydrologic scientists Sultan Alam and Clinton Willson have concluded that although the model data are largely qualitative, the model produces a fairly accurate assessment of alternative diversion project schemes.[52] This emphasis on qualitative observation is critical and represents a shift toward design thinking as opposed to quantitative validation. The model generates information at a scale that may not be the most precise but is particularly useful in developing large-scale concepts.

Hybrid Models

Coastal resilience projects often dissolve the disciplinary boundaries between engineering, ecology, architecture, and landscape architecture. Mirroring this dissolution of disciplinary boundaries, designers adopt the modeling techniques pioneered by engineers to communicate their ideas for water management and coastal resilience. These hybrid models articulate a process rather than a fixed and final form. Echoing the ecological *informe* in time and space, these models combine form and content—topobathy landforms as well as fluctuating water and sediment—to explore the fluid dynamic motion inherent in a system. Although hybrid models may be less precise than complex engineered hydrological models, they test broad-stroke ideas at a moment in the design process where precise measurements would yield little generative information.

Guy Nordenson, Catherine Seavitt, and Adam Yarinsky first used water tank models for *On the Water: Palisade Bay* in 2010. On this estuarine site, where the Hudson River and the East River meet the Atlantic Ocean, the velocity and direction of currents vary dramatically by day and season. The team proposed an archipelago of artificial islands to dissipate wave energy and reduce the water's velocity during storm events. To determine the most successful arrangement of wave-attenuating structures, the team developed a form-finding mechanism from a water tank set atop a light table and connected to a water source and drain. (See Figure 2.35.) For the first trial, doughnut-shaped islands were arranged in a grid. In the second trial, a staggered grid formation with offset rows more successfully slowed the flow of water, creating multiple small eddies. Subsequent experiments used V-shaped forms and tested more complex island configurations. Each array of islands was tested with water and colored dye, simulating directional current flow, eddies, and vortices. Although the flat surface of the light table and the unregulated flow of the water source do not reflect the precision of contemporary hydraulic modeling techniques, the water tank studies generated a strong, graphic presentation of surface speeds and currents, vividly communicating the effectiveness of the island formations in mitigating flow and generating a design logic for various island formations. (See Figure 2.36.)

In her proposal for the 2013 *Yangtze River Delta Project*, Seavitt again used water tank models, this time to test various topographic forms that might control and slow drainage. The project recasts agricultural fields as flood overflow zones by channeling water into a series of poldered fields:

Figure 2.35: A water tank model allows experimentation, facilitating testing of the interaction of new landforms with current, tide, and storm surge.

Guy Nordenson, Catherine Seavitt, and Adam Yarinsky, *On the Water: Palisade Bay*, 2010

low-lying areas of land surrounded by earthen embankments. Although traditional polders are fully enclosed, Seavitt developed the concept of "open" or "passive" polders that function as temporary floodwater reservoirs. Simple clay models were used to examine hydrodynamic flow given changes in the depths of the canals, the angle of canal intersections, and the height, location, and shape of polder perimeters. Submerged in basins of blue water and photographed from above, the models graphically demonstrate the location and extent of water flow. (See Figure 2.37.) Although the topography and the velocity of the water are not precise replicas of the proposed conditions, the models clearly and simply demonstrate the polder forms and the movement of water through them.[53]

Seavitt created additional test models for the Jamaica Bay proposal for *Structures of Coastal Resilience*. In contrast to the water tank models for Palisade Bay and the *Yangtze River Delta Project*, the Jamaica Bay models have a detailed topobathymetric base. The primary site model, a 3-foot by 3-foot continuous-surface topobathymetric rendition of the bay, was CNC milled for precision. Seavitt simulated tidal cycles and surges with pumped dyed water to understand the flow of water through the Rockaway Inlet and within the bay. Residence time at the deep borrow pits of the eastern side of the back bay near John F. Kennedy Airport was observed. By dropping dye into the borrow pits and simulating tidal flow, it was possible to visualize the reduction of circulation and the increase of the residence time of water within the bay, reducing overall water quality. (See Color Plate 5.)

Seavitt experimented with detailed topographic and bathymetric interventions through a set of glycerin working models, cast from inverse laser-cut contour models. Carving into the glycerin topography of Floyd Bennett Field, a former marsh, she tested the form and location of two proposed tidal inlets. The extents of an overwash plain bound by low, earthen berms in Jacob Riis Park on the Rockaway Peninsula were calibrated. Simulating a surge entering through the Rockaway Inlet, Seavitt carved the overwash plain to efficiently allow surge waters to exit the bay and flow back into the Atlantic, sparing adjacent neighborhoods from longer flooding. Lastly, with the Edgemere/JFK Runway Flushing Tunnel Study model, Seavitt tested the effectiveness of below-grade mechanized flushing tunnels to improve water exchange between the ocean and the bay. These model flow experiments were captured on video, documenting the pathways of water movement through the design interventions. The videos assisted in the design process of modifying and adjusting the proposed interventions, but they also provided a clear

Figure 2.36: Time-lapse photographs of water flowing through and around three forms of experimental islands. The islands precipitate trailing eddies, wakes, and vortices, captured in stop motion in the water tank.

Guy Nordenson, Catherine Seavitt, and Adam Yarinsky, *On the Water: Palisade Bay*, 2010

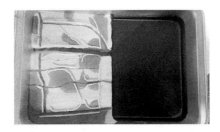

Figure 2.37: Open polder water tank study for the design of the *Yangtze River Delta Project.*

Catherine Seavitt and Guy Nordenson/*Yangtze River Delta Project,* 2013

narrative of how the interventions might perform during surge conditions. In this sense, physical models become strategic and performative design tools rather than instruments of precise data collection. The physical model thus serves to construct ideas and generate a productive approach to working within fluid dynamic systems.

New approaches to coastal resiliency may be investigated by reimagining the representational conventions of plan, section, and topobathy through creative practices that emphasize a transformative *informe*. The ecological formless provides a productive lens through which to view the coast as a dynamic entity. Methods of abstraction, borrowed from both the sciences and the arts, allow novel analysis and flexible design strategies for unknown climate futures. And the use of dynamic modeling processes, both physical and digital, can be reinvented as new design tools, informing decision making at a fluid and gradient coast.

Chapter 3

Reimagining the Floodplain

American geographer Gilbert F. White (1911–2006) presented a compelling argument for a comprehensive approach to managing flood risk within a floodplain region in his persuasive dissertation "Human Adjustment to Floods: A Geographical Approach to the Flood Problem in the United States," written in 1942 and published by the University of Chicago in 1945. White asserted, "Floods are 'acts of God,' but flood losses are largely acts of man."[1] Critiquing a system of reactive legislation that promoted structural engineering solutions to flooding, including seawalls, levees, and dams, White noted that the presence of these structures had provided an overconfident sense of security to the general public and had even encouraged the occupation and development of at-risk flood areas. Throughout his career, White advocated for a more holistic approach consisting of "non-structural" human adjustments to flood risk, supporting the accommodation of flood hazards through the restriction of development within areas that were periodically subjected to flooding, a policy-driven approach of adaptation that he called "floodplain management." Indeed, as engineered flood control structures have failed to prevent disastrous floods in recent decades and as the effects of climate change will produce increased levels of risk, the floodplain management approach is gaining advocates in coastal flood zones. In addition, the ecological components of design are being used to enhance White's policy approach to floodplain management, bringing new and innovative design responses to the development of resilient coasts.

Whereas the previous chapter examined the possibilities of visualizing the coast not as a static line but as a site of dynamic exchange, this chapter explores innovative design methods for intervening within that malleable territory of the floodplain to reduce damage and impacts from both storm surge and incremental sea level rise. Presenting an argument for urban climate adaptation that manages coastal flooding rather than preventing it, design projects are presented that illustrate innovative strategies for neighborhoods, cities, and regions to adapt to a changing climate, not by resisting the sea but by carefully and consciously living with water.

Coastal resilience demands methods of representation that capture the dynamism of water. Here, design proposals are presented that move beyond Gilbert White's policy-driven approach toward floodplain management and enrich his nonstructural position through the use of designed ecologies with enhanced performative capacity. These projects draw from dynamic forms of representation to develop creative propositions that work with, not against, the movement of water, sand, sediment, and ecological species. Drawing from scientific research but also artistic practice, the projects described emerge from innovative tools of visualization and analysis. By first considering the waterfront as a site of exchange, these projects propose the transformation of physically hardened waterfronts, vulnerable to flooding, into softer, more complex coastal sites that might better weather future storms.

Critical to this new vision of the waterfront is a move away from traditional forms of flood control and toward a better understanding of floodplain management. This is a semantic distinction with significant physical implications: Flood control seeks to keep water out, whereas floodplain management allows more nuanced strategies of reducing risk from flooding that expand beyond seawalls and levees to include habitat restoration, stormwater management, new building types, and urban planning measures. The projects explored here offer broad examples of new approaches to floodplain management in practice. More specifically, they address coastal flooding from storm surge and sea level rise through the combination of *attenuation*, *protection*, and *planning*, three actions that roughly correspond to offshore, nearshore, and onshore design interventions. Through a combination of the attenuation and dissipation of wave energy, the protection of coastal infrastructure and development, and the considered planning of cities and regions so that coastal land is developed strategically, the coast might be radically transformed.

A shift away from flood control and toward floodplain management may be achieved through a three-part approach of attenuation, protection, and planning. This chapter presents a series of projects that articulate this approach to coastal resilience. Organized from the scale of a single reef or marsh to a large region or river delta, these projects demonstrate a dramatic shift in considering land, water, and the inhabitation of the increasingly watery areas between them, offering new possibilities for coastal resilience.

From Flood Control to Enhanced Floodplain Management

Hurricane Katrina made landfall along the Gulf Coast of Louisiana on August 29, 2005. The storm led to more than 1,800 deaths and more than $100 billion in damages, making it one of the deadliest and costliest natural disasters in American history. The flooding that devastated huge swaths of New Orleans resulted from the widespread failure of an inadequately managed flood control system. The surge breached levees in more than fifty locations, and floodwaters subsequently spread across three-quarters of the city of New Orleans. Katrina highlighted a central paradox in managing flood risk: Improvements to infrastructure designed to prevent frequent but minor flooding increased the risk associated with less frequent but major events. Neighborhoods protected by levees had assumed a level of safety that was quickly overwhelmed by the catastrophic flooding.

Immediate attention focused on the sudden levee failures in New Orleans, but Hurricane Katrina soon revealed the long-term neglect and mismanagement of both urban systems within the flood walls and natural systems outside them. Long-established and spatialized inequities along race and class lines contributed to the disastrous outcomes of Katrina. A disorganized relief effort failed to protect and evacuate the disproportionately African American and low-income residents displaced from flooded neighborhoods.[2] Together with Hurricane Rita, which followed 3 weeks later, Katrina damaged salt marshes all along the Louisiana coast, drawing attention to the ongoing environmental disaster of the rapid loss of wetlands. Low-lying marshland is disappearing at an alarming rate because of sea level rise and land subsidence, the latter of which is accelerated by the extraction of oil and natural gas and the dredging of navigational channels.

When the levee system in New Orleans failed, it revealed the fragility of an entire floodplain. Katrina exposed both the shortcomings of flood control (preventing flooding with hard structures) and the importance of floodplain management (reducing potential losses from flooding across vulnerable territory through both structural and nonstructural measures, including ecological restoration, urban planning, and policy). Coastal wetlands provide a buffer for inland towns and cities; they slow storm surge, attenuate waves, and absorb floodwaters. Their disappearance threatens not only ecological habitats but also the capacity of the levees behind them to protect communities. Urban planning and policy can also help build resilient communities better prepared for flooding when a levee is breached or overtopped. Flood protection cannot be limited to a single line of defense; rather, it should layer sequential mitigation strategies, thickening and multiplying that line over a broadened conception of the coastal edge.

With increasing concerns about sea level rise and storm surge, this transition from flood control to floodplain management is part of both a national and international conversation about planning for climate change while building on institutions and policies that have managed rivers and coastlines throughout the twentieth century—and even longer in some regions. Although institutional structures are slowly changing, it is important to recognize the historical influences that have shaped flood control for decades.

The dependence on levees in New Orleans is the result of a long history of flood control on the Mississippi River, in particular the Flood Control Act of 1928, which charged the U.S. Army Corps of Engineers (USACE) with building levees along the Mississippi and local citizens with maintaining them. The Flood Control Act was a response to the Great Flood of 1927, which breached natural levees from Illinois to the Gulf of Mexico, displacing hundreds of thousands of people and contributing to the Great Migration of six million African Americans to northern cities. Like Katrina, the 1927 flood cannot be understood as merely a natural disaster—its destruction resulted from a flood control system that favored levees over spillways, outlets, diversions, and cutoffs. In his masterful narrative *Rising Tide*, John Barry describes the development of the "levees only" flood control system from the 1870s through the 1920s. Clashes of personalities and politics rather than balanced science, planning, and engineering led to thousands of miles of levees that constrained the river until it could no longer be contained.[3] The Mississippi levees assured safety where there was none; they decreased the potential for damage during normally elevated water levels, but they also increased possibility of massive destruction during catastrophic events.

Although levees are still a significant part of flood management along the Mississippi, the Great Flood and the subsequent 1928 Flood Control Act led to the federal acquisition of floodplains and the construction of spillways, most notably the Bonnet Carré Spillway, which can release Mississippi floodwaters into Lake Pontchartrain before it reaches New Orleans. The Bird's Point Floodway in Missouri and the Morganza Spillway north of Baton Rouge were also authorized in 1928, and both were activated in the recent 2011 Mississippi spring floods. Thus the Great Flood also marked a turning point, beginning a slow transition away from a "levees only" model of flood control and toward controlled flooding. But spillways designed to release rivers in flood stage provide just one form of controlled flooding. The effective management of periodic flooding in urban settings is another example; during the exceptionally high tides of the *acqua alta* in Venice, for example, the city adjusts to the presence of water in the streets and plazas. Controlled flooding can also occur behind levees that are designed to be overtopped; here, levees may minimize the depth of flood inundation but not fully prevent it during severe events, allowing water into a floodplain designed to sustain a certain amount of water. Finally, controlled flooding can also be harnessed to build new land and restore wetlands. Along the Mississippi River delta to the Louisiana Gulf Coast, scientists have been proposing diversions of the river since the late 1960s to bring sediment to the rapidly disappearing coastal wetlands.[4] Both the Caernarvon Freshwater Diversion off the Mississippi River south of New Orleans and the Wax Lake Outlet off the Atchafalaya River have created new land through sediment deposition. Yet more, and more substantial, diversions are needed to combat land loss along the coast.

The possibilities of ecological enhancement as part of floodplain management and the use of natural systems for reducing coastal storm risk have now brought the USACE to the forefront of a significant paradigm shift. The agency is investigating the impact of climate change on coastal risk and the possibilities of incorporating natural systems into coastal storm risk management. Work done at the two major research centers of the USACE, the Engineer Research and Development Center (ERDC) and the Institute for Water Resources (IWR), is supported by regional district investigations into the transformation of current USACE practices.

Natural and Nature-Based Features

One major aspect of this shift away from established flood control practice is the incorporation of what the USACE defines as "natural and nature-based features" (NNBFs) into their spectrum of responses to coastal risk. A 2013 ERDC report titled "Coastal Risk Reduction and Resilience: Using the Full Array of Measures" summarizes the range of responses to coastal storms.[5] The USACE defines four categories of flood risk reduction measures. Natural features are "created and evolve over time through the actions of physical, biological, geologic, and chemical processes operating in nature." Nature-based features "may mimic characteristics of natural features but are created by human design, engineering, and construction to provide specific services such as coastal risk reduction." Nonstructural measures are "complete or partial alternatives to structural measures, including modifications in public policy, management practices, regulatory policy, and pricing policy." Finally, structural measures, including "levees, storm surge barrier gates, seawalls, revetments, groins, and nearshore breakwaters," are designed to "decrease shoreline erosion or reduce coastal risks associated with wave damage or flooding." "Integration" is described as the process by which these different types of measures can be combined to reduce coastal flood risk and increase resilience.[6]

The "full array of measures" of USACE aligns roughly with what others might define as the range from hard or gray infrastructure to soft or green infrastructure. Whereas structural measures fall into the hard or gray infrastructure category, nonstructural measures follow from soft infrastructural systems—nonphysical networks, systems, and institutions. Green infrastructure encompasses both "natural" and "nature-based" features. These categories are not conceived as mutually exclusive but rather as components to be combined and integrated.

USACE is beginning to implement this spectrum of measures. In January 2015, the USACE released the North Atlantic Coast Comprehensive Study (NACCS), commissioned by Congress as a study examining the impact of Hurricane Sandy along the Atlantic seaboard. Emphasizing vulnerable coastal populations, the NACCS evaluated ten states from New Hampshire to Virginia and the District of Columbia to provide a Coastal Storm Risk Management (CSRM) framework for the North Atlantic region. This framework outlines a series of steps from analysis to implementation to adaptive management and supports adaptable measures that can be layered to improve "redundancy, robustness, and resilience."[7] USACE efforts to identify and implement the "full array of measures" necessary to manage floodplains are significant, but the construction of truly resilient and adaptable coastal regions will demand the collaboration of state and municipal agencies, private developers, property owners, planners, and designers. Floodplains do not respect jurisdictional boundaries; therefore, coastal risk management must engage multiple partners. Moreover, planning and design for coastal resilience and adaptation can also improve coastal cities and communities by providing amenities and resources with multiple benefits beyond flood protection.

Toward Floodplain Management

Another regional management policy is the ongoing Louisiana Coastal Protection and Restoration Authority (CPRA). Initiated in 1990 with the congressional enactment of the Coastal Wetlands Planning, Protection, and Restoration Act (CWPPRA), this was the first federal program dedicated to funding coastal restoration for the State of Louisiana through a cost-sharing plan. With the landfall of both Hurricane Katrina and Hurricane Rita along the Gulf Coast in August and September 2005, the necessity of improving both hurricane protection systems and the natural system of protective wetlands for both the region and the nation's well-being gained urgency. In December 2005, the Louisiana Legislature convened to restructure the state's Wetland Conservation and Restoration Authority to form the CPRA.[8]

CPRA created a Coastal Master Plan for the state of Louisiana in 2012 with an update in 2017 that included a predictive model intended for use as a regional-scale and long-term coastal planning and decision-making tool. The master plan includes a combination of approaches, from traditional hard structural shoreline protection to marsh creation, barrier island restoration, and sediment diversion to initiate land building in the delta region. The Coastal Master Plan also combines both ecological restoration and storm protection, recognizing their interdependence. Yet the plan poses a challenge, as it lacks the investment that would be necessary to protect the full coastline. This requires the difficult task of ranking and prioritizing the many worthy projects distributed across the entire coastal corridor.

Internationally, floodplain management also contrasts fundamentally with the historical Dutch flood control system, Delta Works, which relies primarily on structural measures to keep water from entering the low-lying Netherlands. Initiated in the 1950s, the Dutch Delta Works system consists of dams and storm surge barriers, dune construction and enhancement, and an elaborate system of flood defense dikes.[9] The system is intended to be watertight and highly interconnected; the primary dike rings and secondary regional flood defense dikes depend on tightly controlled central planning.

But in the face of climate change, the Dutch are now reconsidering this strategy of keeping water out, relaxing their strict adherence to flood control and moving toward floodplain management efforts. The Netherlands faces not only sea level rise but also land subsidence and an increased annual discharge from its rivers. After devastating floods at the Rhine River delta in 1993 and 1995, the Dutch government proposed the "Room for the River" program, launched in 2007. The project lowers dikes, opens polders, and builds spillways to allow controlled flooding along four rivers: the Rhine, Meuss, Waal, and Ijssel. It also demanded significant buy-in from local farmers and a more inclusive decision-making process than is typical of the centralized Delta Works.

Additionally, Dutch engineers have begun to embrace natural processes in the construction of offshore artificial dunes. The innovative "sand engine" or "sand motor" strategy uses wind, waves, and longshore current to distribute massive quantities of sand placed offshore to build resilient beaches. Over a period of 10 to 20 years, the natural distribution of this sand will create widened beaches and dunes on the coast of Delfland, near The Hague, with the goal of providing an additional layer of protection to the coastal dikes.[10] The sand motor, initiated as a pilot project in 2006, has just completed its tenth anniversary and is significant in its outreach to local populations through innovative arts programming.

These examples of paradigm shifts in coastal protection and planning in the Dutch context share the characteristics of revisionist and creative design thinking in partnership with local agencies and stakeholders. Engagement with the public is critical for the success of these adaptive coastal resiliency programs, as is the consideration of the coast as a shared civic space that provides benefits to communities throughout the year. The design of innovative and resilient coastal systems must both engage the public imagination and capture the interest of local and regional agencies.

Attenuation, Protection, and Planning

The historic emphasis on flood control in the Netherlands and the Mississippi delta region precipitated failures that were dramatic, sudden, and catastrophic, but through a shift to floodplain management such failures might instead be considered productive—indeed, a first step toward recovery. Layered systems of coastal resilience combine strategies that, if implemented independently, are inadequate against flooding. Wetlands can be inundated, berms overtopped, dunes eroded, and evacuation routes blocked. Yet when combined, inadequate barriers create a new form of redundancy; they work together to change how and where flooding occurs while allowing for some areas to get wet. This shift away from a singular "big fix" defense against flooding and toward a layered system of coastal resilience can be realized through the three-part strategy of attenuation, protection, and planning. (See Figure 3.1.)

Comprehensive flood-resistant design should accomplish all of the following: attenuate and dissipate wave energy offshore to reduce impact on nearshore measures; protect residents with flood protection structures such as levees and seawalls as well as nonstructural measures, such as elevation, relocation, and evacuation strategies; and finally, plan for controlled flooding through urban and landscape design and management strategies that acknowledge the natural floodplain. Although the principles of attenuation, protection, and planning are roughly aligned with offshore, nearshore, and upland features, there is flexibility and redundancy in how they can be configured. This presents opportunities for the important role of creative design in partnership with policymakers, informed by the unique ecology of each site.

For example, a wetland may attenuate waves while a revetment protects coastal land and an elevated home minimizes building damage from flooding. But the wetland might be replaced by a reef and the revetment with a seawall. An elevated evacuation route might be more important in some locations than elevated homes. Measures of attenuation, protection, and planning do not, in most cases, fully eliminate the potential for flooding; rather, they minimize the effects flooding may have on human safety, property, environmental health, and community livelihood. A system that accounts for attenuation, protection, and planning relies on active management; its construction is ongoing and evolving, informed by the behavior of coastal ecology and the social structures of coastal communities. Coastal resilience is never completed; it is a process through which floodplains are managed to adapt to higher water levels while, at the same time, building urban vitality through design.

Designing Layered Systems

The authors of this book first explored the principles of attenuation, protection, and planning in the 2010 book *On the Water: Palisade Bay*, a design master plan for the Upper Bay of New York and New Jersey.[11] Developed

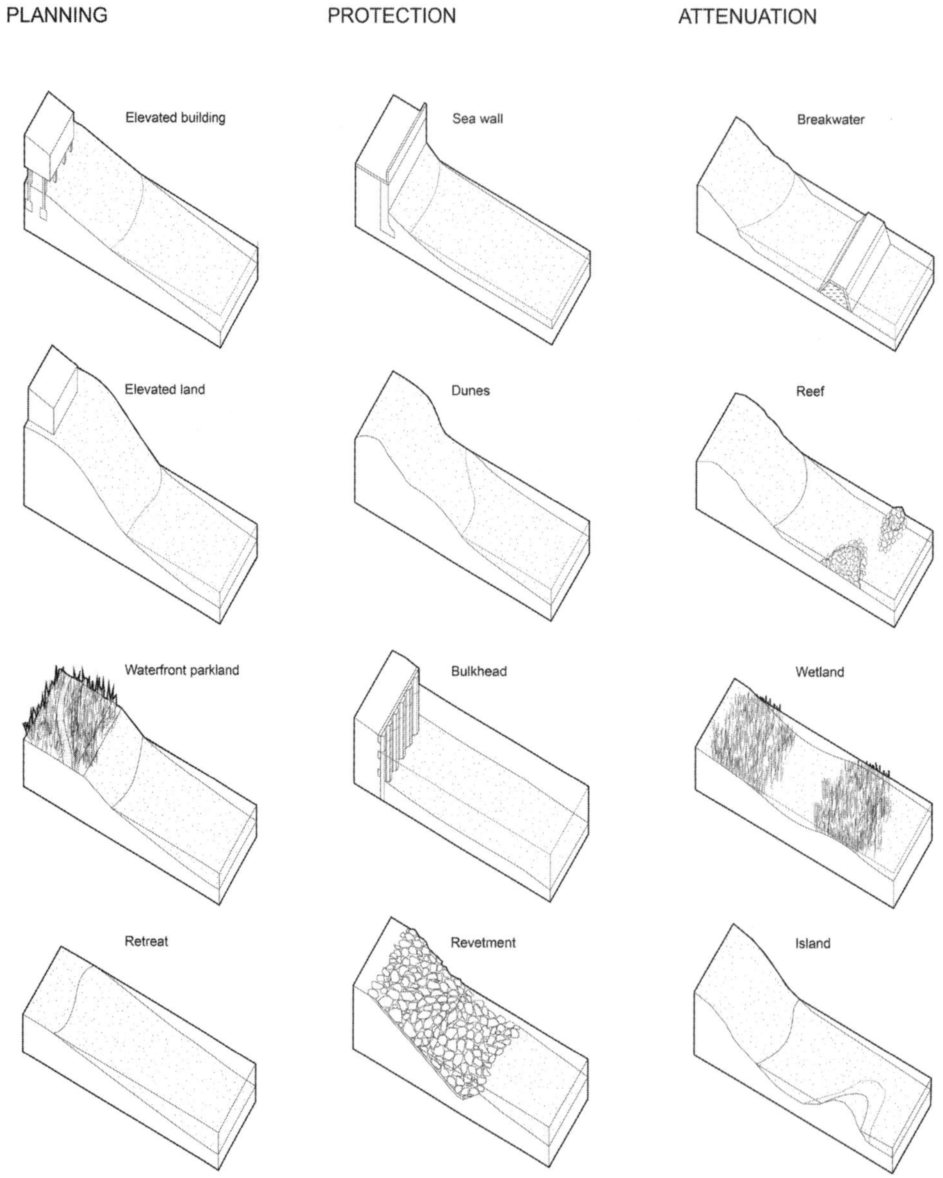

PLANNING   PROTECTION   ATTENUATION

Elevated building

Sea wall

Breakwater

Elevated land

Dunes

Reef

Waterfront parkland

Bulkhead

Wetland

Retreat

Revetment

Island

Figure 3.1: Strategies of planning, protection, and attenuation roughly correlate to upland, coastal, and offshore interventions. When deployed from low to high ground, they provide layered redundancy, critical to resilience.

Catherine Seavitt, Guy Nordenson, Julia Chapman, 2015

in view of sea level rise and the heightened risk of inundation from storm surge, the plan is layered and distributed and offers a conceptual alternative to the more traditional engineered approaches proposed for the region. The work for Palisade Bay responded specifically to a suite of three storm surge barriers proposed to protect the Upper Bay: one placed at the Verrazano Narrows at the entrance to the Lower Bay, a second at the mouth of the Arthur Kill at western Staten Island, and a third at the upper reach of the East River where it meets the Long Island Sound.[12] Instead of closing the Upper Bay to surge from the ocean, as these proposed barriers would do, the plan recommends a series of soft and green infrastructural interventions in and around the bay. (See Figure 3.2.) These interventions are designed to work with existing ecological systems through natural processes to improve environmental

Figure 3.2: Preliminary sketch of typical coastal conditions for the *On the Water* master plan, for the New York/New Jersey Upper Harbor.

Guy Nordenson, Catherine Seavitt, and Adam Yarinsky, *On the Water: Palisade Bay*, 2010

quality, attenuate waves, and protect coastal communities in New York City and New Jersey from severe flooding while creating a new sense of regional urban place centered on the figure of the Upper Bay.

The *On the Water: Palisade Bay* master plan consists of three major elements—islands, piers, and wetlands—supported by secondary elements including oyster racks and reefs, wave and wind turbines, and an artificial reef constructed from decommissioned subway cars. (See Figure 3.3.) New constructed caisson islands at the shallows of the Bay Ridge Flats offshore of Brooklyn and in other shallow areas near Staten Island, lower Manhattan, and the New Jersey shoreline of the Upper Bay would dampen wave energy and provide valuable intertidal habitat for plants, invertebrates, fish, and birds. New and restored historic piers and slips would attenuate waves, improve upland drainage, and store and filter stormwater through reservoirs,

Figure 3.3: *On the Water: Palisade Bay* master plan for the New York/ New Jersey Upper Harbor, showing constructed wetlands, piers and slips, oyster racks, archipelago islands, wave and wind turbines, and an interharbor ferry transportation system.

Guy Nordenson, Catherine Seavitt, and Adam Yarinsky, *On the Water: Palisade Bay*, 2010

bioswales, and permeable surfaces. Restored and constructed wetlands along the New Jersey, Manhattan, and Brooklyn shores would not only attenuate waves to protect inland communities but also create new recreational areas, filter wastewater and runoff, and remediate polluted zones.

As part of *On the Water: Palisade Bay*, the design firm Architecture Research Office (ARO) created a design proposal for the zone around lower Manhattan—one of five zones at the perimeter of the Upper Bay identified in *On the Water* for further detailed study—that transforms the lower Manhattan shoreline to create a new resilient landscape relating to the existing structures (ferry terminals, small buildings, tall buildings, and an elevated highway). The proposal includes an array of barrier islands and breakwater towers just off the tip of Manhattan, designed to slow storm surge and waves while providing habitat for birds and marine life. New wetlands, tidal pools, and breakwaters allow water to slowly traverse the coastline in two directions, improving stormwater retreat and protecting the city from surge. ARO's images of lower Manhattan present a new, greener vision of the dense urban fabric of the financial district. One view juxtaposes the familiar skyline against a new ground: The cityscape is foregrounded by parks and marshland dissolving into the water. (See Color Plate 6.) This image suggests that floodplain management can transform the urban edge, making it more inviting and accessible.

On the Water: Palisade Bay served as the foundation for the organization of a workshop program and exhibition at the Museum of Modern Art (MoMA) in 2010 titled *Rising Currents: Projects for New York's Waterfront*. For the exhibition, ARO elaborated their initial proposal for lower Manhattan, and four other design teams led by New York architects and landscape architects proposed design strategies to mitigate the consequences of sea level rise and storm surge for the four other zones identified in *On the Water*. Less than 2 years after *Rising Currents* was presented at MoMA, Hurricane Sandy struck the East Coast. Storm surge severely affected parts of New York City and the New Jersey and Long Island coasts. Sandy brought national attention to the vulnerability of the region and precipitated new design initiatives, including the Department of Housing and Urban Development's *Rebuild by Design* competition in 2013. Initiated by the authors to complement the USACE North Atlantic Coast Comprehensive Study, *Structures of Coastal Resilience* (SCR) also followed from post-Sandy efforts in 2013 with the objective to investigate how natural and nature-based features might be integrated into comprehensive resilient design systems along the North Atlantic coast. Throughout the 2-year SCR science and design research project, the four university-based SCR design teams met periodically with the USACE North Atlantic Division, ERDC in Vicksburg, Mississippi, and local USACE district offices. These meetings offered opportunities for landscape architects, architects, scientists, and engineers to collaboratively inform one another's work.

In the projects created for *Rising Currents*, *Structures of Coastal Resilience*, and other initiatives, the possibilities of design thinking in collaborative partnership with science and engineering are showcased. Beginning with new analytic visions of the coast, these projects demonstrate the possibilities of building resilience in urban neighborhoods through the integration of coastal attenuation, protection, and planning. The design proposals address the specific flood hazards of each site while creating visually compelling imagery to convey design concepts to broad audiences. Organized by scale and scope, the following proposals range from small-scale interventions into marine ecologies to large-scale regional and deltaic transformations. "Resilient Landscapes" examines projects for New York City and Rhode Island that capitalize on the growth processes of oysters, wetlands, and forests to protect shorelines. "Incremental Adaptation" demonstrates the gradual adjustment of a neighborhood over time to sea level rise and flood risk through improvements to architecture, infrastructure, and recreational space. "High Ground" illustrates the planned development and construction of high and low ground to transform highly vulnerable urban areas. And finally, "Deltaic Flows and Diversions" explores the unique challenges and opportunities of diverting flows of water and sediment toward land building, drainage, and flood protection within riverine deltas. From the scale of a reef to a neighborhood to a city to a delta region, the projects included here demonstrate how layered and integrated systems of floodplain management not only can increase coastal resilience but also can help cities and regions dynamically adapt and thrive in an uncertain future. Creative visualization, design thinking, and a productive collaboration between designers, scientists, and engineers are critical components for the development of successful initiatives for future resilient coasts.

Resilient Landscapes

Through the harnessing, propagation, and management of natural processes, the use of biotic features can develop resilient landscapes that both mitigate floods and enhance local ecologies. Resilient landscapes embrace green infrastructure, and carefully designed landscape features can function as infrastructural systems that engage the indeterminacy of natural systems, guiding processes that yield a range of possible outcomes rather than a single, predictable result. From offshore islands and breakwaters to tidal wetlands to upland vegetated berms, landscape features attenuate, protect, and direct planning efforts below, within, and above the tidal zone. Individual features contribute to surge reduction, but it is only through the combination of interventions working in tandem that landscapes can achieve attenuation, planning, and protection. At the same time, resilient landscapes provide economic, recreational, and flood protection services to adjacent urban communities.

Although green infrastructure is increasingly gaining traction as a viable tool for resilience, the engineering community has been resistant to incorporating these measures into storm protection. Because resilient landscapes integrate multiple approaches rather than depend on a single line of defense against flooding, they are more challenging to test and measure than a hardened seawall or levee. They also require the professional collaboration of different kinds of expertise, from landscape architects and planners to biologists, ecologists, and hydrologists. The design thinking needed to create a resilient landscape must not only draw from art, design, and scientific backgrounds but must also identify creative ways to engage experts and local communities.

Recent projects for New York City demonstrate the integration of resilient landscape strategies with new waterfront initiatives. Brooklyn Bridge Park, designed by Michael Van Valkenburgh Associates, includes the resilient features of wetland vegetation, repurposed maritime piers, and new earthen levees as landscape forms. The East River Waterfront, designed by SHoP Architects and Ken Smith Landscape Architect, features maritime grasses and mussel-attracting habitat atop a storm sewer outfall pipe.[13] West 8's design for Governors Island raises the elevation of the southern half of the island by creating a new topography of hills from demolition debris. The Fresh Kills Park master plan, designed by James Corner Field Operations, repurposes a massive former landfill into a park three times the size of Central Park, spanning the capped landfill, coastal wetlands, and intertidal waterways. Fresh Kills is a long-term project; its construction will continue for decades. These four projects demonstrate the incorporation of resilient features into the design of waterfront parks. But comprehensive proposals for resilient landscapes can move beyond parkland and into urban communities, engaging the dynamic processes of freshwater, ocean tides, and plant and animal ecologies both on land and offshore. Moreover, they can instigate large-scale urban transformations that traverse offshore, nearshore, and upland urban zones.

Reef Attenuation

The design of robust biotic systems often both improves habitat and water quality and reduces the impact of waves and onshore flooding from storm surge. These two objectives are interdependent: The careful management of an ecosystem can contribute to wave attenuation and surge protection, and the investment in ecological forms of hazard reduction can, in turn, support an ecosystem that might not otherwise thrive. The restoration of a bivalve mollusk that once thrived in New York Harbor, the eastern oyster *Crassostrea virginica*, is the foundation of "Oyster-tecture," Kate Orff's project for the 2010 MoMA exhibition *Rising Currents*. Orff and her landscape architecture firm SCAPE proposed an oyster reef at the Bay Ridge Flats, a shallow area of the Upper Bay just south of Red Hook, Brooklyn, that would not only attenuate waves but also improve water quality, create recreational space, and establish the framework for an economically generative aquaculture. (See Color Plate 7.) Orff envisions a spatial network that parallels the life cycle of oysters: Young oysters begin their life in spat tanks and floating upwelling system nursery rafts in a remediated Gowanus Canal before they are moved to a large structure of timber piles and "fuzzy rope" at the Bay Ridge Flats east of Buttermilk Channel and offshore of Red Hook and Sunset Park, Brooklyn. (See Figure 3.4.) This rope structure supports oysters and other marine species, including mussels and eel grass, accreting over time into a reef. As the reef matures, eventually creating a connection northward toward Governors Island, it attenuates waves and mitigates surge in the subtidal and intertidal zones, reducing the impact of wave energy on the shoreline of Red Hook and Sunset Park. The oysters would also filter contaminated water, cleaning the bay and stimulating new economic and recreational opportunities on and around the reef. (See Figure 3.5.)

Orff continues to explore the potential of wave-attenuating reefs to reduce flood damage with SCAPE's project for the Department of Housing and Urban Development's 2013 *Rebuild by Design* (RBD) competition. Her team's visionary entry, "Living Breakwaters," was selected as a winning submission, and the proposal is now fully funded with U.S. Department of Housing and Urban Development Community Development Block Grant–Disaster Recovery (CDBG-DR) funds and is currently being implemented by the New York State Governor's Office of Storm Recovery. "Living Breakwaters" is sited along the

Figure 3.4: Perspectival section at the Gowanus Canal, where a marine economy of oyster production would both improve water quality and support a vibrant space for recreation and industry.

Kate Orff/SCAPE Landscape Architecture PLLC, MoMA *Rising Currents*, 2010

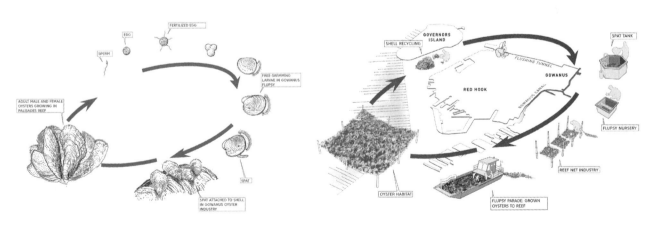

Figure 3.5: The life cycle of the oyster corresponds with activities and industries at specific marine locations near Red Hook, Brooklyn.

Kate Orff/SCAPE Landscape Architecture PLLC, MoMA *Rising Currents*, 2010

south shore of Staten Island on the Raritan Bay at Tottenville, an area significantly affected by Hurricane Sandy in 2012. A resilient layered strategy that stretches from high ground across the shoreline, tidal flats, intertidal zone, and subtidal zone, the project distributes "habitat breakwaters" of specialized concrete at the edge of the tidal flats to reduce erosion, attenuate waves, encourage sedimentation, and provide habitat for plants, fish, and shellfish. Reefs constructed further from shore, in the subtidal zone, encourage habitat growth and protect coastlines from erosion through wave attenuation. SCAPE has explored the incorporation of concrete and mesh structures into the breakwaters to encourage plant and animal inhabitation of this "living infrastructure." Along the shoreline, the project proposes a series of "water hubs" that support a range of water-based recreational, educational, and research opportunities, renewing the Tottenville community's historical connection to the coast. "Living Breakwaters" has been transformed from a visionary proposal into a very real project, one that must address the necessary environmental impact statements, permitting requirements, regulatory approvals, and funding mechanisms for implementation and construction.

Ecological Infrastructure at Jamaica Bay

At a larger scale, resilient landscapes offer an alternative to hard infrastructural solutions in dense, urban, vulnerable communities. Instead of relying on the engineering prowess of a single seawall, resilient landscapes depend on design thinking to strategically combine strategies of attenuation, planning, and protection into an interdependent holistic approach to managing flood risk. Catherine Seavitt's proposal for *Structures of Coastal Resilience*, titled "Shifting Sands: Sedimentary Cycles for Jamaica Bay," defines layers of ecological infrastructure from the barrier island of the Rockaway Peninsula, across the extensive embayment of Jamaica Bay, and at the perimeter shoreline of the surrounding communities at the back bay. The SCR project engages novel strategies of wetland restoration and water quality improvement that are designed to harness natural processes such as tidal flux, current direction, and sediment transport and deposition. Sustaining, restoring, and adapting ecosystems to cope with sea level rise and storm surge will not only enrich habitat and ecologies but also reduce the risk of shoreline erosion and flooding in Jamaica Bay's residential neighborhoods.

Jamaica Bay is a prime example of the confluence of natural resources and urban communities at risk. Located on the southern coastal edge of New

York City and straddling the boroughs of Brooklyn and Queens, Jamaica Bay is a critical and highly managed resource for the New York region. A diverse collection of federal, state, city, and nongovernment entities maintain the bay and its infrastructural systems. The approximately 32-square-mile bay is surrounded by a 142-square-mile watershed, transformed through extensive urbanization into a constructed sewershed under the jurisdiction of the New York City Department of Environmental Protection. The New York District of the USACE is responsible for dredging the western entrance to the bay at the Rockaway Inlet. Much of the land within and adjacent to the bay is publicly owned parkland. The marsh islands at the center of the bay, the western end of the Rockaway Peninsula, and Floyd Bennett Field are all part of the National Park Service's Gateway National Recreation Area, as are the back-bay waterfront parcels south of the Belt Parkway. In 2012, flooding from Hurricane Sandy greatly affected Jamaica Bay and its surrounding neighborhoods, highlighting not only the environmental but also the infrastructural and socioeconomic vulnerability of the area. The widespread damage and slow recovery of the region after Sandy's landfall and the resultant extreme flooding brought the Jamaica Bay region to the forefront of city, state, and federal efforts to improve coastal resiliency. The impact of the storm also demonstrated the imperative for this disparate group of stakeholders to develop a collective strategy to improve the bay's ecological health and increase the disaster resiliency of Jamaica Bay.

Since the 1960s, infrastructure proposals to reduce the risk of flooding from storm surge at Jamaica Bay have explored two primary strategies: a single large-scale barrier for the entire bay or a suite of smaller localized initiatives to protect communities along the perimeter of the bay. A 1964 USACE report proposed a large-scale operable surge barrier at the Rockaway Inlet, at the location of the Marine Parkway–Gil Hodges Memorial Bridge. The 2013 New York City Special Initiative for Rebuilding and Resiliency (SIRR) report proposed a similar barrier further west, at the transect from Manhattan Beach to Breezy Point at the end of the Rockaway Peninsula, thus protecting an even greater area. Other plans take on smaller projects around the bay. The USACE has several ongoing projects along the back-bay inlets, at the marsh islands, and along the beaches of the Rockaway Peninsula.[14]

Seavitt's proposal for Jamaica Bay prioritizes multiple strategic initiatives over a single barrier for two crucial reasons. First, local interventions facilitate adaptive management, whereas future sea level or surge heights might overcome a single barrier. Second, a single barrier is likely to damage the ecological health of the bay, whereas multiple small-scale interventions that incorporate natural features can both enhance ecological health and capitalize on their resilient capacities for storm risk management. The proposed SCR project consists of three integrated strategies: the improvement of water quality and the reduction of flood risk through overwash plains, tidal inlets, and flushing tunnels at the Rockaway Peninsula and Floyd Bennett Field; the strategic verge enhancement of existing high ground at vulnerable back-bay communities; and the restoration of the salt marsh islands in the bay through an "island motor/atoll terrace" that uses minimal quantities of dredged material to create terraced sediment traps at the island perimeters, harnessing the natural process of sediment delivery and deposition. (See Color Plate 8 and Figure 3.6.) The three strategies are interdependent. The flushing tunnels and overwash plains increase an exchange of waters between the ocean and bay,

Figure 3.6: Jamaica Bay Resiliency Plan, consisting of a framework of three design strategies: flow and circulation, verge enhancement, and atoll terrace construction at the salt marsh islands.

Catherine Seavitt/City College of New York, *Structures of Coastal Resilience*, 2015

delivering sediment to the bay that then accretes at the marsh island atoll terraces. These marsh islands, in turn, provide a first line of defense during storms, reducing wind fetch and attenuating wave energy, thus protecting the landscaped verge enhancement at the bay's perimeter during frequent, low-intensity storms. The verge, an elevated and planted berm along the perimeter of the back bay, protects communities but also maintains the ecological health of the bay, whereas the construction of a storm surge barrier at the Rockaway Inlet, a proposal currently under discussion in New York, would potentially jeopardize the ecology and water quality of the bay.

Marsh Island Atoll Terraces

Wetlands provide wave attenuation and erosion protection closer to the shoreline within the intertidal zone—the area exposed during low tide and submerged during high tide—but they are increasingly vulnerable to sea level rise. Encompassing saltwater and freshwater marshes, bogs, fens, and swamps, wetlands offer significant environmental benefits. They slow, absorb, hold, and gradually release floodwater from both storm surge and upland flooding in moderate storms. They break down contaminants and filter polluted water, treat wastewater and storm runoff, and support diverse plant and animal species. Constructed wetlands are often designed specifically for water treatment purposes. In major urban centers such as New York City, where combined sewage infrastructure carries both wastewater and stormwater, rainfall can trigger the direct discharge of combined sewer overflow (CSO) into rivers and bays. Wetlands constructed near CSO outlets could filter contaminated water before it reaches harbors, rivers, and bays. Although strategies for creating wetlands have developed significantly over the last few decades, wetland restoration is not a new idea. In 1878 Frederick Law Olmsted, well-known for his design of New York's Central Park, restored the tidal flows of a stagnant Boston salt marsh in the Back Bay Fens as part of his Emerald Necklace of connected parkland throughout the city. Designed for recreation and water treatment, the Fens also mitigates seasonal flooding.[15]

Sea level rise threatens to submerge the remaining marsh islands at Jamaica Bay. The island motor and atoll terrace provides a strategy for combatting pervasive salt marsh loss within Jamaica Bay. (See Color Plate 9 and Figure 3.7.) A resilient salt marsh ecosystem reduces storm risk to surrounding communities by attenuating waves and reducing wind fetch. The island

Figure 3.7: Contour drawing of the proposed Little Egg Marsh atoll terraces. Established within existing shallow intertidal areas, the terraces increase sediment trapping and deposition at the marsh island's footprint.

Catherine Seavitt/City College of New York, *Structures of Coastal Resilience*, 2015

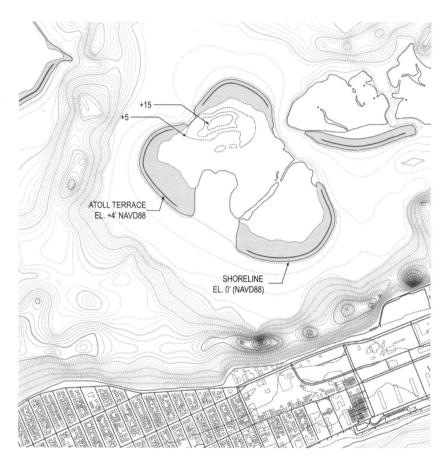

motor maximizes the effects of the strategic placement of dredged material along segments of the perimeter of each marsh island as an elevated atoll terrace. This marsh terrace, with its highest elevation just above the high tide line, provides a gradual slope that encourages the deposition of sediment and the natural upward migration of marsh grass, *Spartina alterniflora*. The initial construction of the atoll terrace thus precipitates a process by which the marsh island accretes both outward and upward. This gradual accretion allows the island to keep up with an incrementally rising tidal zone.

Sediment deposition at the atoll terraces depends on sufficient suspended sediments and the circulation of water within the bay. "Shifting Sands" proposes tidal inlets, overwash plains, and flushing tunnels to improve the water quality and hydrologic flow within Jamaica Bay and from the ocean to the bay. Currently, water and sediment stagnate in the eastern portions of the bay, particularly at Grassy Bay in front of John F. Kennedy International Airport, an area with deep borrow pits. Reducing stagnancy will improve the health of many habitats within the bay and will flush pollutants from sewage and storm runoff out of the bay more quickly. Additionally, tidal inlets, overwash plains, and flushing tunnels would allow surge waters to retreat quickly from the bay, reducing the impact on communities. (See Color Plate 10 and Figures 3.8 and 3.9.)

Verge Enhancement as Protection

While the marsh islands provide a first line of defense against surge, additional layers of low coastal wetlands, earthen berms, and attenuation forests

Figure 3.8: Proposed flushing tunnels at Edgemere, a community at the eastern end of the Rockaway Peninsula. The tunnels would improve water quality at the bay by increasing the bay-to-ocean exchange and allow floodwaters to recede quickly after a surge event.

Catherine Seavitt/City College of New York, *Structures of Coastal Resilience*, 2015

Figure 3.9: Proposal for a marsh inlet at Floyd Bennett Field, a former airfield at the western end of Jamaica Bay. The inlet allows tidal water to circulate through a restored salt marsh.

Catherine Seavitt/City College of New York, *Structures of Coastal Resilience*, 2015

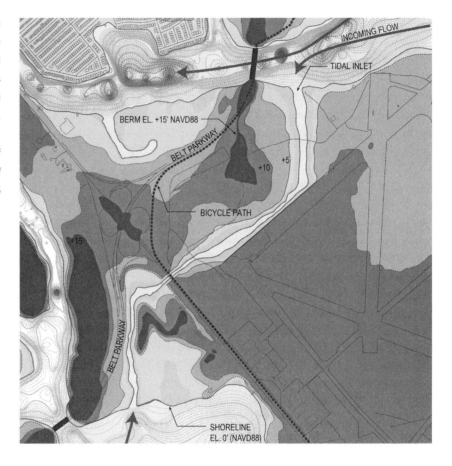

work together to protect vulnerable back bay communities. Together, these layers of elevated vegetation form a "verge enhancement" along an existing infrastructural roadway. (See Figure 3.10.) As new interventions tie into existing high ground at the bridge landings of the Belt Parkway, a continuous elevated buffer zone is developed at the back bay from Mill Creek to Canarsie and Howard Beach. Planted with resilient salt-tolerant species, this new feature also serves as a public space for local residents. Marshes at the bayside of the elevated verge slow incoming waves and surge along with densely planted sunken maritime forests behind the elevated berms, offering communities additional recreational opportunities. Green infrastructural layers protect communities through the resilient qualities of vegetation at the elevated berm while creating new recreational spaces that enhance the urban environment.

Figure 3.10: Proposed verge enhancement at the back-bay neighborhoods of Jamaica Bay. This integrated line of protection is elevated by an earthen berm and merges with the existing infrastructure at the Belt Parkway.

Catherine Seavitt/City College of New York, *Structures of Coastal Resilience*, 2015

Coastal Forests for Narragansett Bay

This strategy of combining subtle topographic change with extensive planting is also used in "Ocean State," another project for *Structures of Coastal Resilience*, led by Michael Van Valkenburgh and Rosetta Elkin with the Harvard University Graduate School of Design. Their proposal for Narragansett Bay, Rhode Island combines large stands of forest vegetation with topographic enhancement and minimal, precisely located, hardened coastal structures. Coastal forests not only attenuate waves and buffer the impact of storm surge on the communities behind them but also reduce erosion. As coastal erosion exacerbates the effects of sea level rise, forests can help maintain habitats and recreational spaces. Elkin and Van Valkenburgh carefully selected specific tree species that are adapted to the saline coastal context. In their proposals for new forests and shrublands, they invoke the ecological principle of disturbance, selecting plant species that will not only survive but thrive in the event of a severe flood or storm. These new forests do not restore a presettlement ideal; rather, they capitalize on natural and human processes to increase the resiliency of both shoreline ecosystems and coastal communities, using the structural capacities of plants as live actors in coastal resilience.

Narragansett Bay is an estuary with a long shoreline vulnerable to coastal flooding. Covering 132 square miles, the bay extends from Providence in the north to Newport in the south. Rhode Island was not severely affected during Hurricane Sandy, but it remains vulnerable to storms. Hurricanes in 1938 and 1954 damaged large areas of the state and led to the creation of the Coastal Resources Management Council (CRMC) in 1971. These hurricanes also led to the construction of the Fox Point Hurricane Barrier on the Providence River, completed in 1966 to protect the city of Providence from storm surge inundation. However, most of the bay lies below this barrier and remains particularly vulnerable to storm surge and sea level rise.

Elkin and Van Valkenburgh's design strategy capitalizes on ecological transition processes already occurring along the Rhode Island coast. The combination of increased urban runoff and higher saltwater levels in the tidal zone threatens marshes and dunes but also allows resilient forest species to migrate closer to the beach front. Plants that tolerate saltwater can colonize precarious coastal zones, helping prevent erosion. The project emphasizes the use of resilient plant species, both native and nonnative, that might be harnessed to create novel forests with extensive root systems that are durable under significant wind and wave stresses. "Ocean State" offers specific visions for three communities within Narragansett Bay, each facing distinct environmental threats. These visions are strategically located to support existing infrastructure in particularly vulnerable locations, including thoroughfares for evacuation during storm events.

The communities of Warren and Barrington, on either side of the Warren River, are vulnerable to both riverine and storm surge flooding funneled up the river from the bay. Here, Van Valkenburgh and Elkin propose pairing berms with planted forests along the river's edge. The coastal forest, bordered on the river side with a low berm and on the inland side with a higher berm, serves as a resilient buffer to the community. The attenuation forest consists of disturbance-adapted plant species that can tolerate inundation during flood events. Particularly resilient plant species that might serve as live actors in this plant-based approach to coastal adaption include sassafras (*Sassafras* spp.), linden (*Tilia* spp.), dogwood (*Cornus* spp.), and American beech (*Fagus grandifolia*). (See Color Plate 11.)

The area known as the Hummocks or Island Park Cove, in Mid Bay, is a low-lying area with a rapidly increasing population, making the area more vulnerable each year. Small summer bungalows now house permanent residents, many of whom depend on saltwater fishing and aquaculture. This growth has overwhelmed the existing sewers and contributed to deteriorating water quality in the surrounding area, threatening livelihoods and precipitating wetland loss. Route 138, a critical evacuation road for Aquidneck Island, also runs through the community. As marshes to the south of the highway disappear and migrate north, this evacuation route becomes more and more vulnerable to storm surge flooding. Elkin and Van Valkenburgh propose a new zone of shrubland south of Route 138, constructed on rolling dunes or hummocks. A shelterbelt of trees between this zone and the road will further protect the road from both surge and wind. The new barrier forest and shrubland will not only reduce erosion but also help the marshland attenuate waves. Staghorn sumac (*Rhus typhina*) is suggested as a resilient shrub species; with pruning, it will grow at a rapid pace, preventing erosion and stabilizing the ground. (See Figures 3.11 and 3.12.)

The community of Sachuest, in the Lower Bay on the southern shore of Aquidneck Island near Newport, is threatened by coastal erosion. Barely 5 feet above sea level, Sachuest is located at the conflux of the Rhode Island Sound and the Sakonnet River. Two large freshwater reservoirs located behind the dunes are vulnerable to saltwater intrusion during a storm event, or even during high tides if erosion and sea level rise continue at the current

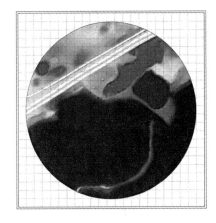

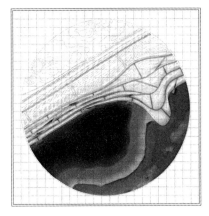

Figure 3.11: Plan of existing and proposed conditions adjacent to an evacuation route near the Hummocks, Rhode Island. The proposed infill provides varied topographic shifts and a field of trees and shrubs that would attenuate wave action during storm events.

Michael Van Valkenburgh and Rosetta Elkin/Harvard University Graduate School of Design, *Structures of Coastal Resilience*, 2015

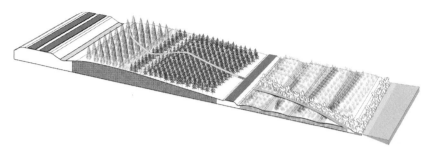

Figure 3.12: Axonometric transect of the Hummocks, showing the planted shelterbelts, establishing publicly accessible pathways and floodable access routes.

Michael Van Valkenburgh and Rosetta Elkin/Harvard University Graduate School of Design, *Structures of Coastal Resilience*, 2015

rate. These reservoirs, Nelson Pond and Gardiner Pond, were salt marshes before they were drained in the late nineteenth century to store drinking water. "Ocean State" proposes creating new reservoirs further inland and transforming the existing reservoirs into a coastal forest. The new forest would slow incoming surge and wind, protecting the reservoirs and communities behind it while creating recreational space and an extension of the nearby Norman Bird Sanctuary and Sachuest Point National Wildlife Refuge. (See Figures 3.13 and 3.14.)

The projects described above integrate natural features such as plant material, biotic habitat, and landforms with existing tidal flows, current patterns, and weather patterns. They exhibit integrative design thinking, demonstrating the use of natural features and systems for design performance, by establishing and propagating plant material, transforming topography and bathymetry, and projecting variable future outcomes through an understanding of the systems that will affect both landforms and biota over time. The preceding examples present the possibility of creating novel landscape systems from a careful reading of a site; the model is no longer a restoration process that seeks to reestablish a past natural condition but rather a vision that projects a future state that can function as a healthy and dynamic ecosystem for a variable future climate. Resilient landscapes perform as infrastructure, mitigating storm risk and improving water quality. Moreover, they provide vibrant public spaces, connecting communities with the coast.

Incremental Adaptation

The development of resilient landscapes with natural features such as forests, wetlands, and reefs is most applicable on public, undeveloped, or preserved land. In many cases these landscapes buffer adjacent residential communities from the effects of storm surge flooding. However, many coastal communities are so low-lying in relation to sea level that additional flood protection measures are necessary. Because of rising sea levels, chronic tidal inundation, and an increased risk of flooding from storm surge, these neighborhoods must adapt in order to thrive. Political leadership continues to debate the merits of the relocation of entire coastal communities to higher ground, out of harm's way. This strategy of managed retreat, often driven by repeated floods, is politically and socially contentious and difficult. But as sea levels continue to rise and chronic "nuisance" flooding becomes common, communities must face difficult choices. Local engineered flood protection often creates a sense of false security and is not always the wisest strategy given the extreme cost of storm protection for future climate scenarios. Adaptation, particularly the strategic design of incremental adaptation, becomes a viable possibility for communities where engineered barriers would be inadequate or cost prohibitive and retreat is implausible.

Though seductive in principle, adapting a vulnerable community in place to manage the effects of sea level rise and storm surge requires creative design thinking. In this scenario, flood control—keeping water out entirely—not only is fallible but also can be dangerous to residents, particularly in the case of an overtopped or breached barrier. Floodplain management offers an alternative approach. Although attenuation and protection are always important, in very low-lying communities the component of planning is critical.

The Amphibious Suburb

For *Structures of Coastal Resilience*, Paul Lewis and his research team at Princeton University's School of Architecture developed a plan for transforming Chelsea Heights, an economically and racially diverse suburban community located at the back bay of Atlantic City, New Jersey into an "Amphibious Suburb." A series of sectional shifts—lifting homes, elevating roads, and digging canals—enables a community to embrace water and benefit from the ecological resiliency of wetlands. Moreover, the project outlines how the neighborhood might adapt over decades to accommodate increasing flood risk, through combined public and private investment. Though designed specifically for Chelsea Heights, the "Amphibious Suburb" is also generic; the design strategies proposed here can be applied to other vulnerable communities.

The dense urban fabric of Atlantic City sits on the low-lying Absecon Island, one of a series of barrier islands along the New Jersey coast. The city is both geographically and economically vulnerable. The recent closure of four casinos in 2014 was the latest setback in the last half-century of economic

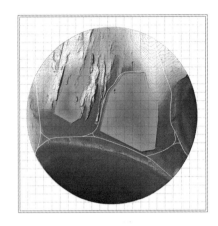

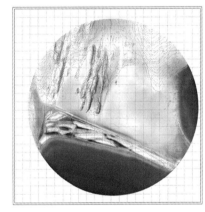

Figure 3.13: In Sachuest, Rhode Island, a constructed freshwater reservoir used for drinking water lies at the margins of the Atlantic Ocean, vulnerable to both storms and saltwater intrusion. By relocating this critical resource to higher ground, the site can be transformed as a planted coastal defense forest.

Michael Van Valkenburgh and Rosetta Elkin/Harvard University Graduate School of Design, *Structures of Coastal Resilience*, 2015

Figure 3.14: Axonometric transect of Sachuest, showing the proposed coastal forest and transformed landscape.

Michael Van Valkenburgh and Rosetta Elkin/Harvard University Graduate School of Design, *Structures of Coastal Resilience,* 2015

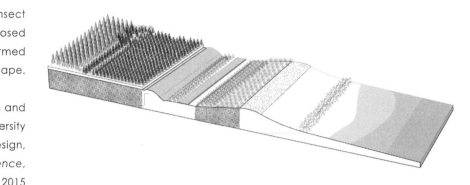

decline. With a per capita income of $18,850, a quarter of Atlantic City residents live below the poverty line, and the city's unemployment rate is one of the highest in the country.

Storms have repeatedly devastated Atlantic City, particularly neighborhoods such as Chelsea Heights that face Absecon Bay. An 1889 hurricane destroyed the Atlantic City boardwalk and flooded seaside property and streets. The 1944 Great Atlantic Hurricane caused severe damage to the city and the New Jersey coast. More recently, Hurricane Sandy caused severe flooding on the New Jersey barrier islands. While the news media paid particular attention to the boardwalk and casinos on the oceanfront side of Atlantic City, some of the worst storm surge flooding occurred at the bayside residential neighborhoods on the western side of the city. While USACE beach nourishment and dune maintenance programs on the ocean side of the barrier island protected development behind them, significant storm surge entered the bay behind Atlantic City through inlets on both sides of Absecon Island, flooding wetlands and back-bay neighborhoods. Water poured through gaps in a discontinuous line of bulkheads and revetments, inundating Chelsea Heights and the adjacent Bader Field with 4 feet of water.

Fortification or Retreat?

The fate of Chelsea Heights and other coastal communities during Hurricane Sandy prompted a familiar debate: Should government agencies fortify communities against future storm surge risk with higher bulkheads and seawalls, or should they facilitate a planned retreat from low-lying neighborhoods where homes should arguably never have been built in the first place? Both options—fortification and retreat—come with high financial and social costs. Building bulkheads to protect Chelsea Heights could cost between $24 and $36 million.[16] High bulkheads also visually disconnect the neighborhood from the water and damage the ecology of surrounding wetlands. If these defensive walls were to fail during a storm event, the resulting damage to buildings behind them could be catastrophic. A planned retreat from the neighborhood would reduce storm risk and allow the restoration of back-bay wetlands. However, retreat would be expensive and would exacerbate the urban decline plaguing Atlantic City, reducing the tax base necessary for revival. If the government were to buy the 608 homes in Chelsea Heights at their average value of $227,000, the total cost would exceed $138 million and would lead to more than $3 million in annual lost tax revenue.[17] Neither fortification nor retreat is ideal, but residents are also in a particular bind: Many want to leave but cannot sell their homes.

Lewis proposes the "Amphibious Suburb" for Chelsea Heights as an alternative to both fortification and retreat. It provides the benefits of each option—wetland restoration and protection from storm surge—by retrofitting the neighborhood to accept water during both daily tidal cycles and storms. The "Amphibious Suburb" is a sponge, designed to protect the neighborhoods of Chelsea Heights and the rest of Atlantic City beyond. The transition would be gradual, drawing on public investment as well as private initiative and using both nonstructural strategies and nature-based features. The result is a more resilient neighborhood that provides an improved quality of life for residents and increases property values. (See Color Plate 12.)

Adaptation through Vertical Transformation

The "Amphibious Suburb" depends on three vertical transformations—the lifting of existing and new homes, the elevation of existing roads, and the depression and conversion of back alleyways into canals. (See Figure 3.15.) As the risk of flood damage and associated insurance rates increase, homeowners will be encouraged to raise their homes above the base flood elevation (BFE) designated by the Federal Emergency Management Agency (FEMA). Abandoned lots, in combination with new canals, will encourage the migration of wetlands into the neighborhood, around and below the lifted homes. During moderate flood events, the canals and wetlands will reduce wave action and absorb floodwater. Elevated roads will provide access to all homes and safe evacuation routes in the event of a severe storm. Moreover, as a series of barriers, the roads will slow storm surge as it moves from the bay toward the rest of Atlantic City. A new elevated road that also functions as a berm will run along the north side

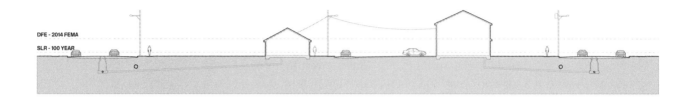

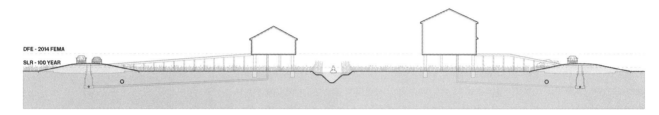

Figure 3.15: The "Amphibious Suburb" at Chelsea Heights is transformed through discrete vertical shifts, including lifted roads and houses, liberating space at the ground plane for wetlands to migrate through the neighborhood.

Paul Lewis/Princeton University School of Architecture, *Structures of Coastal Resilience*, 2015

of the neighborhood, buffering surge while allowing water into the neighborhood canals and wetlands through a series of culverts. The space between the new perimeter road and the existing West End Avenue will become a wetland park, a new neighborhood amenity connecting residents to the water.

The adaptation of Chelsea Heights will be implemented incrementally. (See Figure 3.16.) The neighborhood's capacity to manage storm surge will increase as sea level rises and the risk of severe storms heightens over the century. Initial construction of the elevated road and berm on the north side of the neighborhood will provide immediate protection from storm surge and a much-needed evacuation route. Culverts through the berm will begin the process of canal construction and wetland migration. By 2050, the canals, wetland park, and elevated roads will be complete. Many homes will be lifted, and wetlands will begin to permeate the neighborhood, replacing lawns and abandoned lots. By 2100, Chelsea Heights will be fully amphibious, with wetlands emergent throughout the community.

Although Lewis's proposal is designed specifically for Chelsea Heights, the resilient features within it—wetlands, lifted homes, elevated roads, and canals—could be introduced into other similar back-bay communities along the New Jersey coast. Like Absecon Island, most New Jersey barrier islands feature sprawling suburbs constructed in the 1950s and 1960s, before coastal wetlands were protected from development in the early 1970s. With sea level rise and changes in storm frequency and intensity, almost all barrier island communities fall within FEMA's 100-year floodplain. Some of these suburban communities already feature canals, as do other coastal neighborhoods in California, Florida, and Long Island. Although canal communities were designed to provide water access and views to residents, not for resilience or adaptation, these are now ideal sites for incremental coastal adaptation. (See Figure 3.17.)

House Lifting

The amphibious suburb depends on nonstructural measures such as zoning regulations and insurance policies to encourage or require property owners to lift their homes. After Hurricane Sandy, many homeowners took advantage of available grants to lift their existing homes onto piles. Other homeowners who decided not to rebuild outfitted their empty lots with a lifted foundation in order to sell the land. House lifting is reactionary, as contractors speedily adapt existing structures or home models to be elevated on piles. Lewis and his team documented the unique architectural qualities of homes in New Jersey coastal communities and produced a set of axonometric drawings that chronicle the unusual architectural elements that homeowners and contractors use to mitigate the gap between ground and house. One home features a "surrogate lawn," a large, wraparound deck lifted to the level of the home. In another, a raised parking pad positions the homeowner's car between the ground level and the house, yielding a long ramped path from the street to the front door. In a third, the front door is dropped to the ground level, concealing an interior stair up to the house. Multiple stairs and decks flank these houses. These drawings inventory various makeshift solutions, but they also highlight the unique architectural problems that arise from lifting a house nearly a full story above the ground.

Lewis's analysis of the architectural issues that result from the lifting of single-family homes in flood-prone neighborhoods documents a new informal vernacular building type. Yet these makeshift architectonic transformations invite the possibility of developing new architectural thinking and novel designs for resilient coastal structures. Single-family homes may indeed be raised, but urban townhouses or larger multifamily apartment buildings in vulnerable coastal regions pose particular challenges. Many Atlantic City townhouses have been abandoned, leading to urban blight. In their project for *Rising Currents*, the nARCHITECTS team led by Mimi Hoang and Eric Bunge develop a new architectural solution. Rather than lifting homes off the ground,

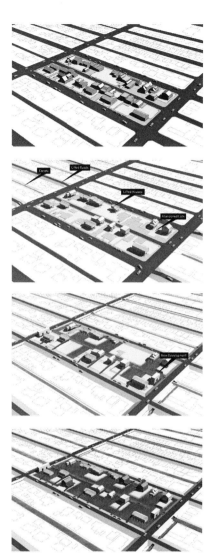

Figure 3.16: A typical block in Chelsea Heights, with phased adaptation in 2014, 2030, 2070, and 2090. Canals replace alleys, with all local traffic moved to newly elevated roads. The combination of new canals, lifted homes, and open lots would encourage the migration of protective wetlands into Chelsea Heights.

Paul Lewis/Princeton University School of Architecture, *Structures of Coastal Resilience*, 2015

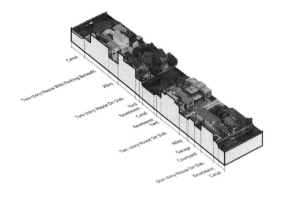

VENICE, CALIFORNIA

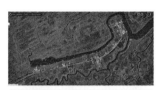

GALLOWAY TOWNSHIP, NJ

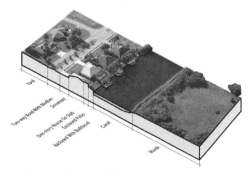

CAPE CORAL, FLORIDA

Figure 3.17: Existing canal communities in Venice, California; Galloway Township, New Jersey; and Cape Coral, Florida. The amphibious suburb references existing neighborhoods with integrated water features as recreational amenities and neighborhood assets.

Paul Lewis/Princeton University School of Architecture, *Structures of Coastal Resilience*, 2015

nARCHITECTS proposes hanging homes from a steel superstructure. (See Figure 3.18.) Their project, "New Aqueous City," imagines transformed water-based neighborhoods on the shoreline of low-lying Sunset Park in Brooklyn and the eastern shore of Staten Island. Here, a shared bridge structure would support townhouses constructed from lightweight materials. Hung from a uniform datum, each townhouse would extend toward sea level below, the height of its lowest floor determined by how much risk the building owner chooses to assume. The bridge structure would maintain density in an urban site.

Incremental adaptation is an innovative approach to resiliency, addressing the sequential transformation of at-risk communities over time. Traditional engineered flood protection infrastructure assumes a singular design flood elevation—an assumption that does not consider changing climates, rising sea levels, or more frequent and stronger storms. Incremental adaptation, as demonstrated by Lewis's "Amphibious Suburb," provides a method by which cities and neighborhoods can adapt to new sea levels and storm surge hazards through multiple efforts, from architecture to infrastructure to ecological restoration, which may occur simultaneously or in sequence over time. And the invention of new architectural solutions—for single-family homes, townhouses, and multifamily apartments—is a necessary component of both incremental adaptation and new urban futures.

Figure 3.18: "New Aqueous City," a proposal for inverted coastal housing and floating wetlands in Sunset Park, Brooklyn.

Mimi Hoang and Eric Bunge / nARCHITECTS, MoMA *Rising Currents*, 2010

High Ground

In an amphibious suburb, houses rise, roads elevate, and alleyways sink to bring water in and out of the neighborhood safely. At the scale of a suburban neighborhood, these vertical shifts will provide significant change. But the logic of lifting and sinking, of elevating and depressing, can also translate to a larger urban scale. Topographic transformation is not without precedent. For centuries, even millennia, before European exploration, Native Americans built large burial and ceremonial mounds in the Mississippi and Ohio River valleys; some theories identify this topographic construction as a form of flood protection. After the devastating 1900 hurricane struck Galveston, Texas, the entire city—all 500 blocks—was lifted more than 16 feet. Because tsunamis—huge tidal waves produced by underwater earthquakes—can inundate coastal areas with little warning, the construction of high ground or vertical evacuation centers is critical to saving lives in tsunami-prone regions. For example, Stanford University and the organization Geohazards International are currently prototyping tsunami evacuation parks—human-made hills with recreational facilities—in Padang, Indonesia.[18]

Reshaping the coast at an urban or regional scale through cut and fill is a dramatic design move, but it can significantly reduce flood risk. In essence, this strategy suggests reorienting flood barriers so that, when no longer parallel to the coast, they do not prevent water from moving inland but rather encourage it to flow inland in strategic locations. Although the process of constructing high ground requires large-scale planning, areas of high ground can also attenuate wave energy at their lowest points, protecting important infrastructure farther inland. Topographic transformation is just the first step in a long process of adaptation as people, plants, and animals migrate and adjust to a more varied terrain and a longer, more crenellated shoreline. Because this strategy encourages gradual adaptation to rising sea levels and accommodates a sudden influx of surge waters, it offers a prime example of the use of floodplain management to serve vulnerable coastal areas.

The Crenellated Coast

In their project for the MoMA *Rising Currents* exhibition, Paul Lewis, Marc Tsurumaki, and David Lewis of LTL Architects reimagine the vast, flat, and vulnerable Jersey City shoreline as a topographically varied urban park. (See Figure 3.19.) Most of Liberty State Park lies just 5 feet or less above mean sea

Figure 3.19: Satellite image collage of the New York–New Jersey Upper Harbor showing the crenellated coastline of the proposed Water Proving Ground at Liberty State Park in Jersey City, New Jersey.

Paul Lewis, Marc Tsurumaki, David J. Lewis/LTL Architects, MoMA *Rising Currents*, 2010

level. Without intervention, sea level rise will gradually inundate this low-lying parkland, owned and operated by the National Park Service and the state of New Jersey. Titled "Water Proving Ground," the project proposes a rigorous cut-and-fill operation that transforms the precarious flood zone of Liberty State Park, Ellis Island, and Liberty Island into an urban amenity programmed for use during both high and low water levels. New high and low terrains yield an exciting exchange of tourism, aquaculture, recreation, and research. "Water Proving Ground" continues a historical legacy of shaping topography; this part of the New Jersey coast was built on landfill between 1860 and 1928. Through the construction of new high ground, a historic site might once again serve as a critical point of exchange in the New York Harbor.

LTL Architects use diagrams to demonstrate how a sequence of design operations can transform a 5-mile coastline into a crenellated 45-mile edge to protect both the immediate region and the communities beyond. (See Figure 3.20.) The design team proposes redistributing existing fill by carving out water channels between four elevated "fingers" or piers. A cross-grain then cuts through and between these four peninsulas, further carving the

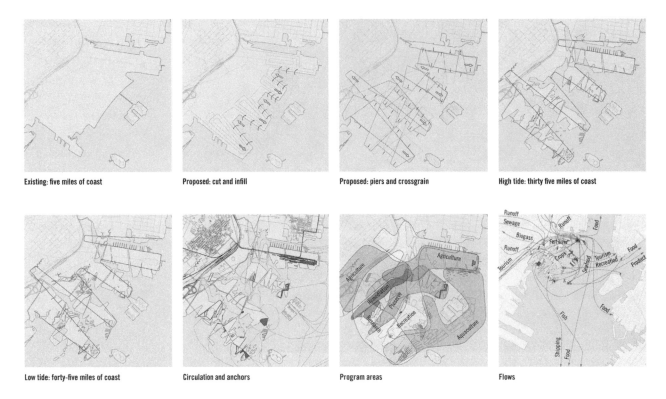

Existing: five miles of coast

Proposed: cut and infill

Proposed: piers and crossgrain

High tide: thirty five miles of coast

Low tide: forty-five miles of coast

Circulation and anchors

Program areas

Flows

Figure 3.20: Diagrams of proposed cut-and-fill strategy transforming the low-lying Liberty State Park into a series of piers and slips, programmed to accommodate the flows of people, goods, and water.

Paul Lewis, Marc Tsurumaki, David J. Lewis/LTL Architects, MoMA *Rising Currents*, 2010

land and increasing the length of the coastline to slow the pace of flooding. The result is a series of triangular spaces that vary in elevation; some capture and hold water during high tides or flood conditions. By contrast, on a flat site storm surge moves inland like a wall of water, quickly inundating vast territories.

These sculpted fingers accommodate programmatic uses calibrated to the shifting dynamics of water. Each triangular segment of land is programmed with agriculture, aquaculture, research, or recreation, depending on the relationship of the topography to tidal flow. A fish hatchery, a hydrological testing center, a tidal park, and a "water lodge"—a recreational center complete with swimming pools, kayak storage, and access to trails and campsites—all inhabit the low-lying and flood-prone zones where the park extends into the

Upper Bay. (See Color Plate 13.) Here, changes in water levels facilitate environmental, research, educational, and recreational activities. An amphitheater with a floating stage creates a flexible performance space. Agricultural plots, a produce market, a wastewater treatment plant, and playing fields are arrayed on higher ground, less vulnerable to frequent water level fluctuations but equipped to absorb flooding during low-frequency, high-impact events. As water rises, flooding the low-lying land, the spine and key access routes of the four piers remain protected.

Fingers of High Ground

"Water Proving Ground" encompasses only about 5 square miles of publicly owned land, but similar cut-and-fill operations can transform much larger urban sites. Landscape architects Anuradha Mathur and Dilip da Cunha, together with their team at the University of Pennsylvania School of Design, propose a strategy of "fingers of high ground" for the low-lying and subsiding city of Norfolk, Virginia. Their project for *Structures of Coastal Resilience*, titled "Turning the Frontier," draws inspiration from the existing crenellated coastal morphology of the Tidewater region of Virginia along the Chesapeake Bay. The alternation of high and low ground occurs at a small scale as well. Along the James River and at its conflux with Chesapeake Bay, for example, marshy terraces are adjacent to areas of firmer, higher ground. At Plum Tree Island National Wildlife Refuge, on the bay near the city of Poquoson, thin forested ridges are striated with low marshlands.

Situated at the mouth of the Chesapeake Bay and the convergence of the James and Elizabeth Rivers, Norfolk is vitally connected to water. It is also one of the American cities most vulnerable to the effects of sea level rise, compounded by local land subsidence. Since the 1940s, land in the Southern Chesapeake region has been subsiding at a rate of 1.1 to 4.8 millimeters per year, as a result of both localized groundwater pumping and global postglacial adjustments of the earth's crust.[19] Over the last 80 years, mean sea level in Norfolk has risen 14.5 inches and is predicted to rise between 23 and 62 inches over the next century.[20] Several areas of the city experience flooding during extreme high tides multiple times each year. In addition, the many creeks and streams within the city overflow with heavy rainfall during storms, causing upland flooding. Norfolk is the site of critical national infrastructure, including the largest naval station in the world and an important commercial port.

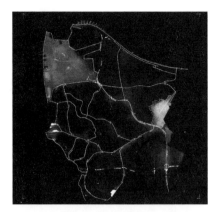

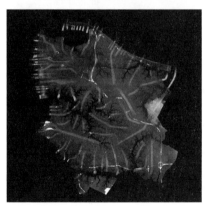

Figure 3.21: Unlike a system closed by levees and gates that require "completion" in order to be effective (top), an open system of "fingers of high ground" creates a series of alternating high and low grounds.

Anuradha Mathur and Dilip da Cunha/University of Pennsylvania, *Structures of Coastal Resilience*, 2015

The city of Norfolk has proposed the construction of walls, gates, and pumps in various locations, but these measures do not offer long-term solutions to prevent flooding given sea level rise and climate change.[21] Rising sea levels combined with the increasing intensity of storms will overwhelm a system designed to keep water out and large areas dry. Instead, Mathur and da Cunha radically propose to "turn the coast" of Norfolk, so that the land does not meet the water along a front—a beach, dune, or seawall—but rather through a series of perpendicular, elevated "fingers of high ground." Like "Water Proving Ground," this proposal would lengthen the coastline significantly, allowing it to wind inland through inlets that amplify and mimic the morphology of streams, creeks, and rills that already punctuate the Virginia coastline along the Chesapeake Bay. This crenellated coastline protects developed land on higher ground from both riverine and surge flooding, as it slows water moving in both directions and provides areas to capture and hold excessive floodwater. A system of strips of high ground, perpendicular to the coast, would be gradually implemented, helping the region avoid catastrophic flooding. Extensive dredging operations in the rivers and bay near Norfolk can provide material for building high ground in strategic locations.

Mathur and da Cunha compare a closed system of levees and floodgates, similar to that proposed by the city, with their proposal for fingers of high ground. (See Figure 3.21.) In the illustration of the city's closed system, the narrow levees and flood walls stand out as white elevated lines enclosing a series of isolated, lowland sections of the city, rendered in black. Not only is the city divided into segments, but the survival of each segment depends solely on the strength of the wall surrounding it. In the illustration of Mathur and da Cunha's open system of fingers of high ground, the city becomes much more continuous and gray, lightly punctuated by strips of white and black, high and low ground. Floodable zones are far more ambiguous, but this is seen as an advantage; some areas will remain dry while others adapt to a more amphibious condition. The continuity of low ground is critical to ensure postflood drainage.

The placement and design of fingers of high ground respond to existing conditions. Fingers of high ground can be built around and adjacent to elevated highways and rail lines. Existing ridges can be widened and extended. Coastal forests with varied topography, such as loblolly pine hummocks, can absorb additional ground material. Urban voids, abandoned lots, and the land beneath elevated infrastructure can become areas where flooding is expected and even encouraged. Lower-lying land can be designed to accommodate flexible programs that could withstand a flood event, including urban agricultural plots and athletic fields.

Once constructed, fingers of high ground not only provide elevated land for short-term refuge and long-term resettlement but also provide ecologically sensitive infrastructure that may facilitate natural processes of migration and adaptation. These narrow strips of land transform as they stretch inland. A pier in the bay may transition into a forest and finally into ground for emergency facilities or housing. A low-lying wetland between two fingers may become a grassy park and then agricultural land. Just as human inhabitants of Norfolk will move to higher ground as the risk of flooding increases, plant and animal species will also migrate inland and upland as their habitats shift or disappear. Fingers of high ground facilitate that transition yet ensure that it happens slowly, allowing plants and animals to migrate along gradients of water salinity, water depth, soil composition, humidity, and air temperature.

Although the proposal envisions a new topography for the entire city of Norfolk, Mathur and da Cunha illustrate detailed master plans for two particular fingers of high ground: Lambert's Point and Willoughby Spit. For each of these sites, they outline construction phasing over the next 80 years. Through illustrated sections, they convey the integration of flood management with other urban and ecosystem services through variations in the relationship between high and low ground along each finger.

Mathur and da Cunha's proposal for high ground at Lambert's Point not only offers flood protection and mitigation but also facilitates the site's transition, already nascent, to postindustrial use. Lambert's Point is located on the Elizabeth River southwest of Old Dominion University and northwest of central Norfolk. Home to a major coal-exporting facility, a wastewater treatment plant, and a golf course, the site is characterized by an expansive industrial rail yard and long shipping piers. The rail lines enter the yard from a ridge that extends inland and offers an initial spine for a finger of high ground. As a flood mitigation strategy paired with a major urban redevelopment corridor, Lambert's Point has the potential to become a major public asset for the city of Norfolk. (See Figure 3.22.)

The phased development of Lambert's Point that begins along the shipping piers extends inland along the rail line ridge. A renewed pier and a restored wetland together form a living shoreline. They support the existing wastewater treatment facility, processing overflow waste through biotic systems while also attenuating waves during storm conditions. A wetland is also restored alongside the tracks at the base of the rail yard; here, water is

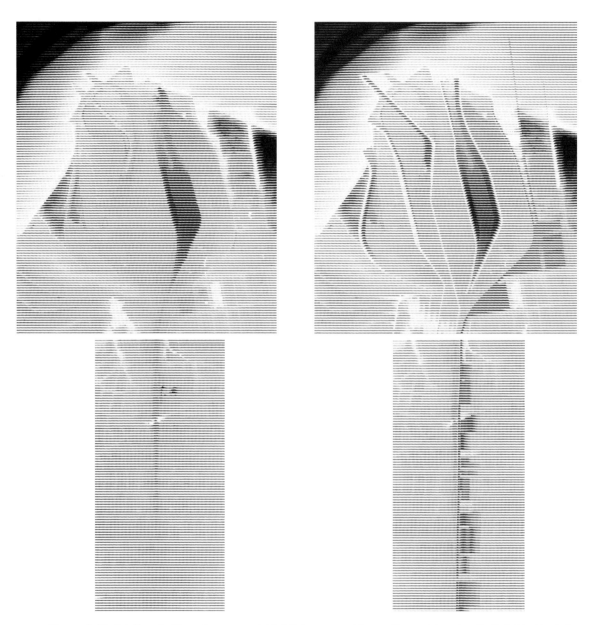

Figure 3.22: Existing (left) and proposed (right) topographic sections at Lambert's Point in Norfolk, Virginia. The proposed topography creates a ridge of high ground that extends inland, providing a spine for transportation, industry, and mixed-use development.

Anuradha Mathur and Dilip da Cunha/University of Pennsylvania, *Structures of Coastal Resilience*, 2015

held and then slowly released into the river. Along the ridge, plots of land are raised incrementally and strategically alongside the northern edge of the tracks. Elevated land is then slated for development while adjacent lower plots of land are transitioned into rainwater holding and treatment basins. A public walkway and bike path runs along this elevated ridge, creating a vegetative buffer between coal-laden trains and urban neighborhoods.

At Willoughby Spit at the northern tip of Norfolk, another finger of high ground builds on an existing raised highway, U.S. Route 64. New lateral fingers extend off of this main spine toward the water. These lateral strips of elevated land break waves and slow storm surge. Critical buildings, including a school located on the site, can be relocated to zones of higher elevation, and the low grounds between the lateral fingers become sites for recreation, urban agriculture, and stormwater collection. A field of piles not only attenuates waves but also provides a desirable habitat for fish and plants. (See Figure 3.23.)

When the coast is understood as a frontier or a barrier dividing land from sea, the processes that must cross that line—tides, plant and animal migrations, drainage—are stilted. Rotating that line perpendicular to the coast creates a more resilient city. By "turning the coast" of Norfolk through fingers of high ground, Mathur and da Cunha are also effectively "turning the frontier," acknowledging the continuum of a historic process begun in Virginia. English settlers came to Virginia in 1607, encountering the Atlantic coast as their first frontier; it was the first of many as Europeans moved westward across the continent over the next three centuries.

Both LTL Architects' "Water Proving Ground" and Mather and da Cunha's "Turning the Frontier" explore the possibilities of raising land elevation as a resilient strategy for the future occupation of coastal territories. They subvert the notion of the elevated seawall as a protective structure parallel to the coast by creating pierlike structures that operate perpendicularly to the shore. By letting water in, rather than futilely attempting to keep it out, as well as strategically constructing new higher grounds, a new approach to an urbanism that embraces water may be developed, one that is more agile and performative in adapting to change.

Deltaic Flows and Diversions

In addition to Norfolk, Mathur and da Cunha have explored other coastal cities and river deltas, from Mumbai to the Mississippi. In the 1990s they led an extensive research project documenting the shifting terrain of the

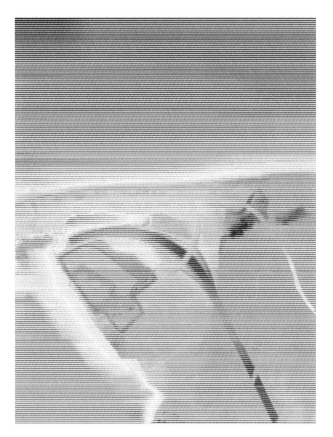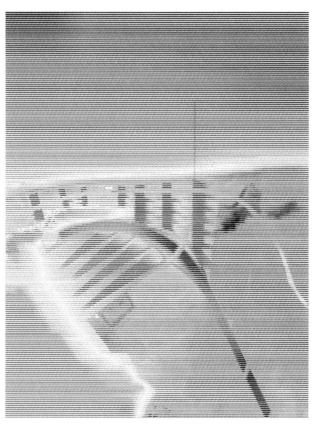

Figure 3.23: Existing (left) and proposed (right) topographic sections at Willoughby Spit in Norfolk, Virginia. The proposed topography creates elevated ground along Route 64 and a series of wetlands and channels extending toward the ocean.

Anuradha Mathur and Dilip da Cunha/University of Pennsylvania, *Structures of Coastal Resilience*, 2015

Mississippi River and the many attempts to control the river's natural meander. Their work was motivated by the 1993, 1995, and 1998 Mississippi floods and resulted in an exhibition and book titled *Mississippi Floods: Designing a Shifting Landscape*, published in 2001.[22] Through drawings, maps, collages, and text, Mathur and da Cunha demonstrate the instability and dynamism of this mighty river and the land it traverses. Rivers and their deltas challenge an understanding of solid ground, changing topography and geography at tremendous scales. In its journey to sea level, a river not only carves through territory but also builds land, carrying and eventually depositing sediment in the delta region. This process makes riverine and deltaic systems not only important to the formation of topography but also a site and source of coastal resilience.

Although deltaic environments can be particularly vulnerable to sea level rise and storm surge, as well as upland flooding, they also provide unique opportunities to harness the river's resources of sedimentary deposits to enhance resilience. In an estuarine river delta, water flows in two directions. While the river's freshwater drains with gravity into the sea, the tides push water inland. Unique ecosystems develop in this zone of interchange between freshwater and saltwater. Rivers carry sediment as they meander across the floodplain. When the river turns and slows, sediment suspended in the water is deposited, building new land along riverbanks and in deltaic plains. Where the river's meander has been contained within a channel, sediment is moved along the channel unless obstructed or diverted. But when diverted strategically, the suspended sediment in the water can be harnessed to build new land in vulnerable, low-lying locations. As the deposited sediment accretes and forms new land, wetland marshes and natural levees are formed that can protect inland communities.

River deltas are some of the largest sites of dynamic exchange on the planet. They vividly illustrate that the coast is not a line on the horizon but rather a thick system that extends both inland and offshore, facilitating the movement of water, sand, sediment, animal and plant species, commerce, and people. This complexity makes the estuarine delta a rich site for design thinking, in need of integrated approaches that consider the natural, social, political, and cultural contexts that drive change. Here, historical perspective is particularly important to design thinking; river deltas have been controlled by industrial and engineering structures for centuries, and they must be understood as the confluence of both human and hydrological narratives.

Mississippi River Diversions

With wetland loss and land subsidence occurring at an alarming rate, the Mississippi delta has become a critical location for research into river diversions. (See Figure 3.24.) Louisiana State University coastal scientists Robert Twilley and Clinton Willson propose five freshwater diversions of the Mississippi River to deliver and deposit sediment along a vast stretch of the delta. Two diversions at Wax Lake and Terrebonne draw water and sediment from the Atchafalaya River. Two diversions at Bayou Lafourche and Davis Pond draw from the Mississippi River north of New Orleans. The fifth and final

Figure 3.24: Mississippi River delta wetland loss between 1990 and 2010. Wetlands are disappearing at a rate of 25 square miles per year.

Michael Blum and Harry Roberts, Coastal Sustainability Studio, Louisiana State University, 2010

diversion follows the Mississippi River gulf outlet and Bayou la Loutre to deposit sediment east of New Orleans near Lake Pontchartrain. One of these diversions, at the Wax Lake delta outlet, is currently in operation. Here, the Atchafalaya River, whose diverted flow from the Mississippi is controlled by the Old River Control Structure, deposits sediment through an artificial channel constructed by USACE in 1973, building new ground as much of the rest of the region subsides.

Currently, the Old River Control Structure north of the Mississippi–Louisiana state border diverts 30 percent of the Mississippi's flow down the Atchafalaya River; the remaining 70 percent flows through the main channel of the lower Mississippi River past New Orleans to the Gulf. This 30/70 ratio is mandated by the U.S. Congress. Without the control structure, the Mississippi River would likely change course, as it has many times before, gradually shifting its main stream to the Atchafalaya River channel. Twilley and Willson propose that more water and sediment should travel down the Atchafalaya as well as through additional diversions. Instead of adhering to a constant flow distribution throughout the year, flow distribution percentages could vary depending on the overall flood rate and season to increase sediment deposition in multiple locations across the delta.[23]

In 2010, a Princeton research team led by Guy Nordenson and Catherine Seavitt collaborated with the Coastal Sustainability Studio at Louisiana State University, led by Jori Erdman, Lynne Carter, Jeffrey Carney, and Elizabeth Mossop, to visualize Twilley and Willson's proposal for five large-scale Mississippi River diversions of water and sediment. A series of controlled flooding diagrams illustrates the incremental shift of flow volumes over a period of 50 years. (See Figure 3.25 and Color Plate 14.) Collages made from aerial photographs depict the new wetland terrain that might be rebuilt through each diversion. A large physical topographic/bathymetric model of the Mississippi delta also showed the relationships between land and water in the region.

Twilley, Willson, and the Coastal Sustainability Studio continue their investigation into Mississippi diversions. All were involved in the *Changing Course* competition, an independent competition led by local and national leaders in academic, industrial, commercial, and philanthropic institutions intended to build off the Louisiana Coastal Master Plan. The 2015 competition sought integrated proposals for using the Mississippi River's water and sediment to rebuild coastal wetlands in the Mississippi delta region below New Orleans. The project also engaged design teams with local communities that would be affected by flooding and sediment deposition from the proposed diversions at the Mississippi River delta to envision new futures for the region.

Yangtze River Delta

China's Yangtze River delta exhibits many of the same characteristics of the Mississippi delta and faces some of the same challenges. The Yangtze travels over 3,900 miles, draining 20 percent of China's land area. It carries a massive volume of water and an enormous supply of sediment, depositing 600 million tons of mud and silt each year into the East China Sea. The Yangtze delta region, including Shanghai, is barely above sea level and highly vulnerable to storm surge and sea level rise. The land is also subsiding as groundwater is pumped out from the alluvial soil to support extensive urban development. Hard-engineered levees, dikes, and seawalls surround the urban areas, but the effects of climate change continue to threaten their effectiveness. The Chinese are investing billions of dollars in seawalls along the East China Sea coast. Dubbed the new "Great Seawall," this barrier claims more land for development but also threatens coastal wetlands and their ecosystems.[24]

Figure 3.25: Controlled flooding flow diagrams for proposed diversions at the Mississippi River delta, Louisiana.

Catherine Seavitt and Guy Nordenson/Princeton University with Louisiana State University's Coastal Sustainability Studio, 2010

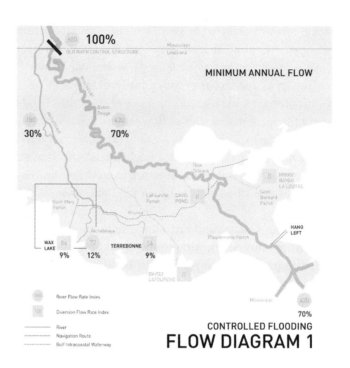

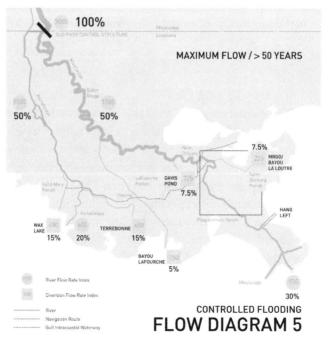

In a project exhibited in Shanghai at the Princeton–Fung Global Forum "The Future of the City" in January 2013 and at the "Shanghai Urban Space and Art Season" in October 2017, Catherine Seavitt and Guy Nordenson propose a new system of floodwater management for the Yangtze River delta. Drawing on historic Chinese techniques, Seavitt and Nordenson's *Yangtze River Delta Project* (YRDP) addresses both coastal flood protection and upland drainage. (See Figure 3.26.) The YRDP was inspired by the flood management strategy of Yu the Great (2200–2100 BCE), founder of the Xia Dynasty, who created a system of low earthen irrigation canals to draw riverine floodwater into agricultural fields. The system relieved pressure on the river and slowed the flow of floodwater to the sea. Moreover, the agricultural fields, mostly rice paddies, benefitted from the gravity-based irrigation system. Rice paddy irrigation techniques such as terracing and bunding—the construction of low earthen ridges to protect stepped terraces—are still used today. Reinterpreting Yu's technique, Seavitt and Nordenson designed a system of "passive" polders for the contemporary Yangtze River delta region. A fully enclosed polder demands mechanized pumping to remain dry, and these polders tend to subside, requiring further pumping during storm events. A "passive" or "open" polder, by contrast, does not fully enclose a territory but allows water to drain out into the surrounding hydrologic watershed. It holds water temporarily during a flood and allows it to retreat slowly through gravity. The YRDP research group explored this gravity-based retention and release strategy with physical water tank models. The entire region of the urban metropolis of Shanghai, its surrounding agricultural fields, and the coastline of the East China Sea are addressed as a massive water holding and releasing network, allowing the region to confront the issues of surge flooding from typhoons with a broad strategy that moves well beyond the singular, but fallible, Great Seawall of China.

Designing a Resilient Coast

Gilbert F. White's lifelong advocacy for a holistic approach to floodplain management remains a significant precedent for new design approaches to coastal resiliency. In addition to White's insistence on smart policy management of floodplain land use, a collaborative and layered design approach that incorporates the ecological performance of natural systems will also support and enhance a resilient coast. Through both innovative design thinking and a comprehensive strategy of attenuation, protection, and planning, the creative

Figure 3.26: Site model of the *Yangtze River Delta Project*, showing the open polders and chenier ridges that recall the historical agricultural landscape of the region. These new features would reduce the vulnerability of the region to flooding from typhoon storm surge.

Catherine Seavitt and Guy Nordenson, *Yangtze River Delta Project*, 2013

work of designers, scientists, and engineers can support coastal communities as they face the impacts of climate change, sea level rise, and the increased risk of flooding and storm surge.

Structures of coastal resilience operate at many scales. Resilience is supported by diverse and layered techniques, including novel plant and animal ecologies, new architectural designs, large-scale land building, floodwater capture and release, and deltaic sediment delivery. These are the new coastal infrastructures, layered with ecological and bioengineered functionality while supporting new vibrant and adaptive urban communities at the coast.

Chapter 4

Mapping Coastal Futures

When a hurricane makes landfall it brings extreme wind, heavy rain, powerful waves, and storm surge. Of these hazards, storm surge is often the most dangerous: Historically, surge has caused more deaths than wind damage. During a hurricane, wind exerts stress on the water, causing currents that push water downwind. The surface of the sea can rise suddenly and dramatically rise as wind speeds approach their maximum.[1] Water can rush rapidly into zones of shallow bathymetry and quickly flood the land. With climate change, storm surge will become more threatening. Sea level rise will elevate mean tide levels, increasing the height of storm surge by simple numerical addition. At the same time, hurricanes may become more intense, and the more intense storms more frequent, also contributing to the chances that severe surge will lead to flood inundation in coastal communities.

The projects described in the previous chapter mitigate the effects of sea level rise and storm surge by restructuring the built environment. These design visions assert that architects, landscape architects, and urban planners should design for resilience and adaptation at vulnerable coastal locations, acknowledging and preparing for the indeterminacies of disturbance events. Accordingly, the planning authorities of states and municipalities should consider coastal adaptation and resilience when regulating land use, encouraging or discouraging development, managing public lands, and envisioning the future of their communities. This approach to coastal resilience and adaptation can be described as a planning-for-hazards model. Hazards are defined as the physical impacts of a storm event: flood extents, storm surge strength, and water velocity.

Risks, by contrast, are the consequences of those hazards. A risk-and-replacement model of flood management differs strategically from a planning-for-hazards model. In a risk-and-replacement model, the financial value of physical assets, and potential damage to those assets, drives policy. The cost of a protective measure—flood insurance or a seawall, for example—is weighed against the cost of replacing the assets it might protect. A risk-and-replacement model operates outside adaptation; it assumes that whatever is destroyed by flooding will be replaced in precisely the same manner. The National Flood Insurance Program (NFIP), one of the United States' primary flood management tools, uses a risk-and-replacement model to manage flood risk across the nation. U.S. Army Corps of Engineers (USACE) projects have also historically followed a risk-and-replacement model, determining the height of a seawall through consideration of the value of vulnerable property behind it. This book, by contrast, advocates a planning-for-hazards model that incorporates urban and regional planning alongside local protection measures. A planning-for-hazards model emphasizes the health and safety of both people and ecosystems over the protection of existing financial assets and considers how resilience and renewal might drive new planning efforts rather than returning to a previously existing condition.

A planning-for-hazards approach to coastal resilience depends on the rigorous analysis and mapping of the hazards themselves, independent of the financial risk they might precipitate. This means evaluating the chances of flooding from storm surge in a local spatial context and then mapping those chances onto a city or region. Hazard maps identify the geographic areas most vulnerable to flooding in the event of a hurricane, as well as the depth of the flooding to be expected as a factor of the severity of the storm. Like the representational tools described in Chapter 2, the mapping tools and techniques described in this chapter are selected and developed to be useful for design at an urban and regional scale rather than providing flood hazard information at the scale of an individual property.

Because hurricane events are unpredictable, mapping storm surge depends on probabilistic analysis. Scientists cannot predict what one unique storm *will* look, like so they must instead evaluate what a large range of possible storms *could* look like and the likelihood of the flood outcomes. Design maps are charged with the task of presenting substantive information about the future, informed by a limited selection of past events and a selection of highly variable environmental factors. Counterintuitively, probability offers decision makers a measure of control over the effects of an indeterminate system. In his 1990 intellectual history *The Taming of Chance*, philosopher of science Ian Hacking traces the development of two parallel narratives during the nineteenth century: first, the emergence of probability and statistics as a dominant mode of inquiry; and second, the fundamental shift from determinism to indeterminism in the natural and social sciences. Hacking argues that these two narratives are interrelated. Although probability and statistics initially explained underlying causal laws, eventually they came to reveal "laws of chance," exposing the inherent indeterminism of systems. Hacking makes a critical point about this intellectual transformation: "There is a seeming paradox: the more the indeterminism, the more the control."[2] In other words, indeterminism considers the chance of multiple effects, and probability offers a reasoned means to determine the likelihood and severity of those effects. In the social and natural sciences and also in engineering, probability enables decision making and risk taking when outcomes are not clear. In the context of storm surge hazard mitigation, a probabilistic assessment of surge allows property owners, communities, and agencies to proactively address flooding to reduce property damage and prevent the loss of life.

Despite the ubiquity of probability in contemporary life—from economics to medicine to business to climate—it remains difficult to represent. Mapping probabilistic information poses a particular challenge. Maps typically define territories with discrete edges; they designate spaces within a municipality or a state and those outside its borders. However, probabilistic information suggests ranges and spectrums, conveying multiple possibilities rather than, for example, distinctly delineated flood zones. It calls for new forms of mapping that illustrate indeterminate scenarios instead of determinate predictions.

In the fluid context of coastal resilience, sometimes abstraction yields clarity. Often, the precision of information concerning the chances of floodwaters reaching a certain height is less important than how that information is embedded into cultural understanding. In Japan, stone tablets known as "tsunami stones" are scattered along the coastline, marking the extent of tsunami inundation from as many as six hundred years ago. (See Figure 4.1.) These inscribed stones not only commemorate lost lives but also warn future generations to seek higher ground.[3] Tsunami stones embed a flood map into the territory. A tool of remembrance, tsunami stones reveal what would otherwise be forgotten. Like a high-water mark on a beach, they register a dynamic history. In order to be effective tools of communication, maps also require design thinking, especially when they must convey events that are impermanent, fleeting, and abstract.

This chapter argues for the production and use of design maps as a critical tool for developing and evaluating coastal resilience design proposals. Through the analysis of existing flood maps and a detailed description of the mapmaking process used in the *Structures of Coastal Resilience* (SCR) project, it provides a roadmap for scientists and designers to develop flood maps for future projects. While acknowledging the complexities of sea level rise and storm surge modeling with the many variables at work in climate change, this chapter calls for proactive interdisciplinary work to map flood hazards so that communities might take action in building resilience at the scale of their community. The chapter begins by identifying the difference between navigational charts, created for the purpose of moving through an

Figure 4.1: A stone tablet in Aneyoshi, Japan, one of hundreds arrayed along the coast, with an inscription warning residents of the extent of seismic-induced tidal waves. Many of the stones are more than six centuries old.

Photo © 2011 Ko Sasaki/*The New York Times*/Redux

existing territory, and flood maps, designed to provide hazard information for planning and design. Analysis of the current status quo in flood mapping in the United States, the Flood Insurance Rate Maps (FIRMs) produced by the NFIP, demonstrates how these maps function as navigational charts, reaffirming hazardous development in flood-prone areas, and are thus inadequate tools for the design of coastal resilience. The second section presents dynamic hazard mapping, an alternative strategy for producing flood maps. This method takes into account the uncertainty of future flood hazards and the impact of climate change on sea level rise and storm surge. Moreover, the combination of design maps into matrices can present a more complete picture of changing hazards and demonstrates how design interventions might reduce the vulnerability of coastal communities over time. However, this mapmaking process depends on a rigorous probabilistic data analysis that accounts for the dynamic nature of storms in a changing climate. The third section outlines how this analysis of dynamic data was produced for the SCR project, as a prototype for future research. Finally, this chapter examines how probabilistic design maps might allow the adoption of dynamic performance-based design standards into coastal resilience work to allow communities to support design and building for uncertain future climates.

Navigational Charts and Flood Maps

As scientist and philosopher Alfred Korzybski famously wrote in 1931, "The map is not the territory." Korzybski was concerned with semantics, not geography, but his statement has been referenced by scholars in fields ranging from cybernetics to linguistics to media theory. In the disjunction between a map and the territory it references there is slippage, which can be a liability but is also a site of opportunity. The space between a place and its representation allows for agency in that representation. Like other representations, maps can be agents of power, controlling people and land through classification and codification. Historically, two-dimensional world maps have represented imperialist agendas. (Perhaps most famously, the Mercator projection, invented in 1569, though useful for marine navigation, enlarges regions farther from the equator while diminishing colonized regions closer to the equator.) But maps can also be subversive and resistant, illustrating the world in new ways. Not only does the territory inform the maps, but the map can inform the territory.

This chapter is concerned with maps that inform the design and planning of coastal territory for resilience. These maps challenge representational

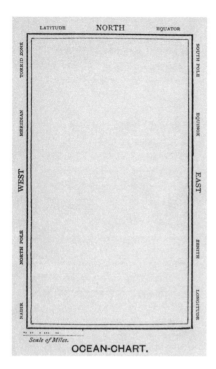

OCEAN-CHART.

Figure 4.2: Henry Holiday, *Ocean-Chart*, from *The Hunting of the Snark*, by Lewis Carroll, 1876

conventions because they must account for how territory will change with climate change, acknowledging shifts in sea level and transformations of the landscape. A distinction can be made here between navigational charts and design maps. Navigational charts guide a navigation of the world as mapped; they depend on the static nature of territory. For a map to be not just a navigational chart but a tool for design, it has to work—and look—differently; it must accommodate changing terrain and a shifting coastline. Design maps must account for the changing nature of nature and the amplified transformation provoked by climate change. Moreover, design maps must accommodate design proposals, illustrating the shifting interface between the transformation and management of coastal regions and the geological and ecological processes that effect and react to those transformations.

Navigational Charts

Henry Holiday's "Ocean-Chart" illustration accompanies Lewis Carroll's 1876 poem "The Hunting of the Snark." (See Figure 4.2.)

A bounded rectangle with cartographic terms circling its edges and a "scale of miles" at the bottom, the "Ocean-Chart" is a blank map, an empty ocean "without the least vestige of land."[4] Carroll mocks the "conventional signs" of cartography that may restrict the imagination of explorers, but the joke is on the crew, who eventually realize that a blank map provides little navigational use:

> He had bought a large map representing the sea,
> Without the least vestige of land:
> And the crew were much pleased when they found it to be
> A map they could all understand. . . .
> "Other maps are such shapes, with their islands and capes!
> But we've got our brave Captain to thank:
> (So the crew would protest) "that he's bought us the best—
> A perfect and absolute blank!"
> —Lewis Carroll, "Fit the Second: The Bellman's Speech,"
> in *The Hunting of the Snark (An Agony in Eight Fits)*

While Holiday's "Ocean-Chart" provides little information, Jorge Luis Borges's Map of the Empire from his paragraph-long short story "On Exactitude in Science" provides far too much. Drawn at a scale of one-to-one, the Map of

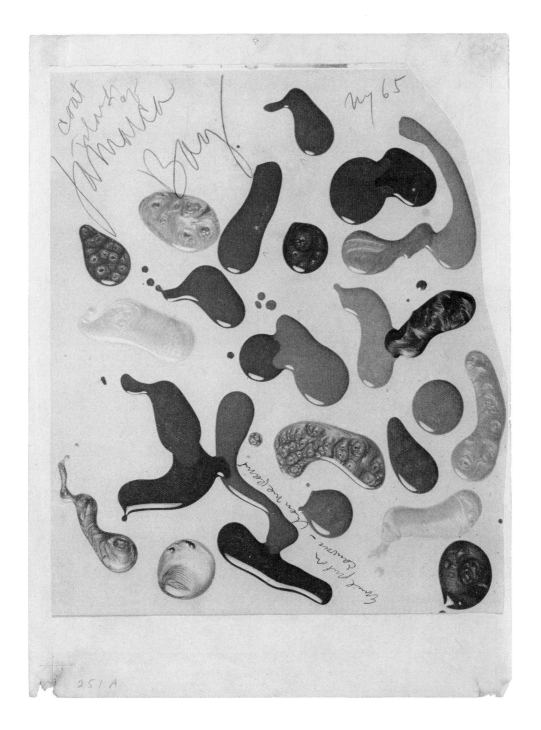

Plate 1: Claes Oldenburg, *Notebook Page: Spills of Nail Polish, "Jamaica Bay."* Collage with ballpoint pen, pencil, and clipping, New York, 1965.

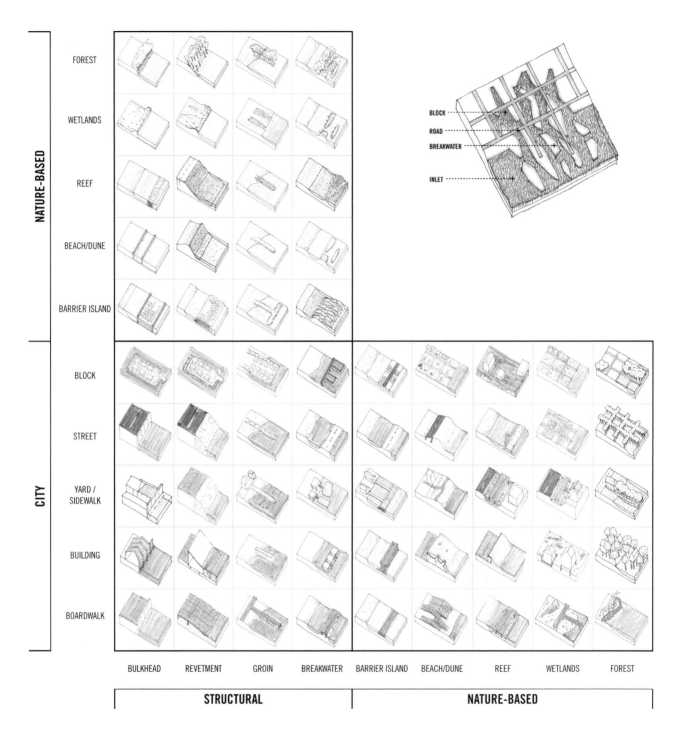

Plate 2: Speculative matrix of hybrid coastal conditions, combining urban, structural, and nature-based infrastructure.

Paul Lewis/Princeton University School of Architecture, *Structures of Coastal Resilience*, 2015

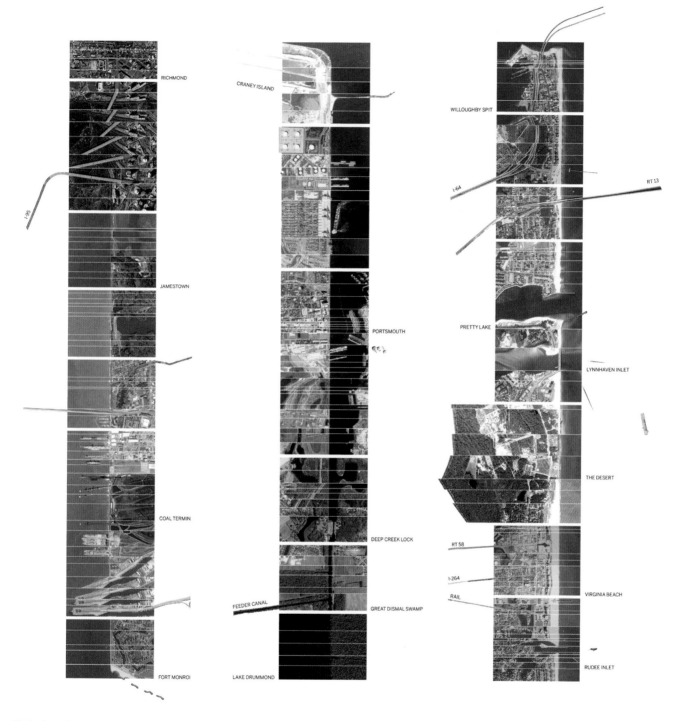

Plate 3: Collaged visualizations of Virginia's Tidewater country. Left to right: Fall Line along the James River from Richmond to Fort Monroe, Swamp Canal from Craney Island to Lake Drummond, Beach Front of Southern Virginia from Norfolk's Willoughby Spit to Virginia Beach.

Anuradha Mathur and Dilip da Cunha/University of Pennsylvania, *Structures of Coastal Resilience*, 2015

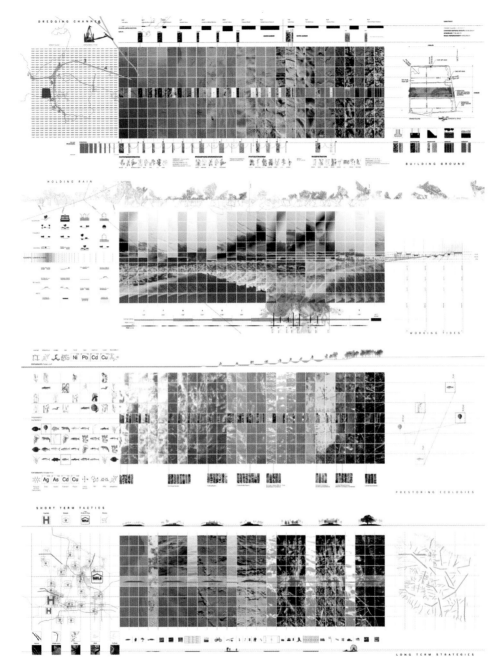

Plate 4: Operational gradients of the lower Chesapeake Bay. Top to bottom: Dredging harbor channels to building high ground; holding rain from upland flooding to accommodating storm surge; processing gray, black, and salt water to "prestoring" new ecologies; developing short-term tactics for managing storms to long-term adaptation strategies.

Anuradha Mathur and Dilip Da Cunha/University of Pennsylvania, *Structures of Coastal Resilience*, 2015

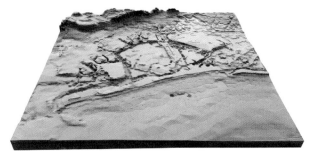

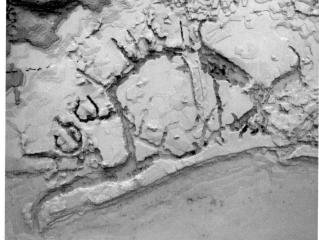

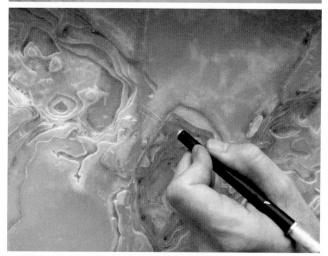

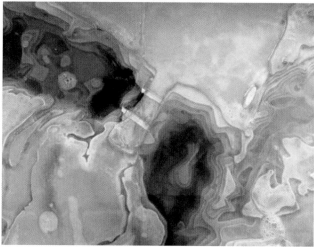

Plate 5: Detailed morphology of Jamaica Bay, from the scale of the watershed to local features within the embayment, is examined through the use of hydraulic water models to provide feedback to the design process.

Catherine Seavitt/City College of New York, *Structures of Coastal Resilience*, 2015

Plate 6: Transformation of the existing seawall of lower Manhattan with a new intertidal zone and offshore islands, revitalizing a wetland ecology and creating a buffer zone for wave attenuation.

Guy Nordenson, Catherine Seavitt, and Adam Yarinsky, *On the Water: Palisade Bay*, 2010

Plate 7: Oyster-tecture reef proposed for the Bay Ridge Flats in New York's Upper Harbor, with floating paths and anchorage areas creating a rich web for recreation and aquaculture while alleviating wave energy.

Kate Orff/SCAPE Landscape Architecture PLLC, MoMA *Rising Currents*, 2010

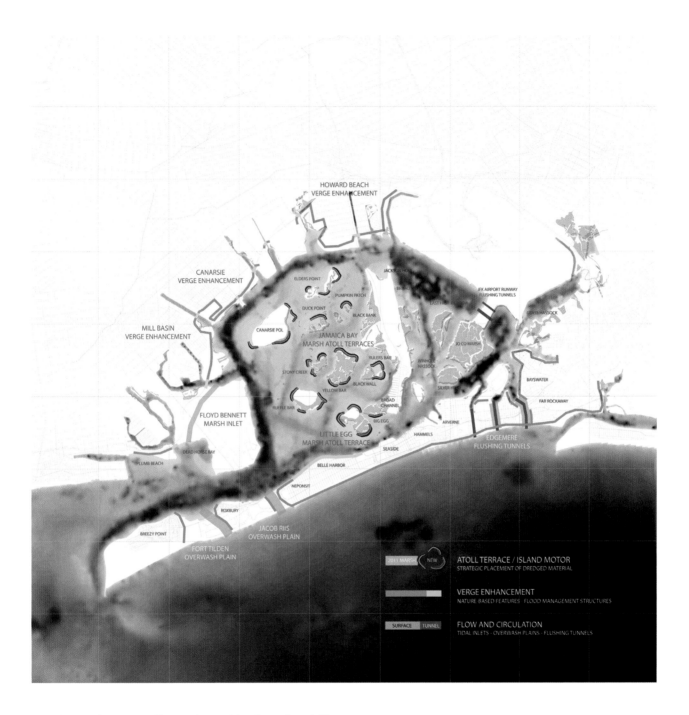

HOWARD BEACH
VERGE ENHANCEMENT

CANARSIE
VERGE ENHANCEMENT

MILL BASIN
VERGE ENHANCEMENT

ELDERS POINT

PUMPKIN PATCH

DUCK POINT

BLACK BANK

CANARSIE POL

JAMAICA BAY
MARSH ATOLL TERRACES

RULERS BAR

STONY CREEK

YELLOW BAR

BLACK WALL

RUFFLE BAR

FLOYD BENNETT
MARSH INLET

BROAD
CHANNEL

LITTLE EGG
MARSH ATOLL TERRACE

BIG EGG

DEAD HORSE BAY

PLUMB BEACH

BELLE HARBOR

NEPONSIT

ROXBURY

BREEZY POINT

JACOB RIIS
OVERWASH PLAIN

FORT TILDEN
OVERWASH PLAIN

JACKSON HOLE

BRANT

EAST HIGH

JFK AIRPORT RUNWAY
FLUSHING TUNNELS

GRASS HASSOCK

JO CO MARSH

WINHOLE
HASSOCK

SILVER HO

BAYSWATER

FAR ROCKAWAY

ARVERNE

HAMMELS

SEASIDE

EDGEMERE
FLUSHING TUNNELS

2011 MARSH | NEW | **ATOLL TERRACE / ISLAND MOTOR**
STRATEGIC PLACEMENT OF DREDGED MATERIAL

VERGE ENHANCEMENT
NATURE BASED FEATURES · FLOOD MANAGEMENT STRUCTURES

SURFACE | TUNNEL | **FLOW AND CIRCULATION**
TIDAL INLETS · OVERWASH PLAINS · FLUSHING TUNNELS

Plate 8: Jamaica Bay Resiliency Plan, addressing vulnerability and coastal storm risk management by merging novel techniques of ecosystem restoration with layered nature-based features.

Catherine Seavitt/City College of New York, *Structures of Coastal Resilience*, 2015

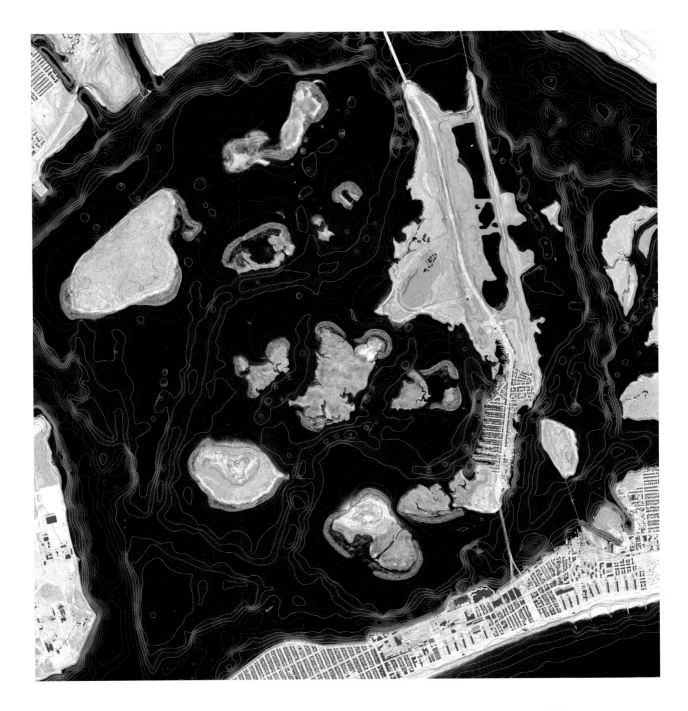

Plate 9: Plan of the proposed atoll terrace design strategy for Jamaica Bay, establishing elevated marsh terraces at the perimeter of marsh island footprints, allowing sediment deposition and accretion through the activation of an "island motor," driven by currents and tide.

Catherine Seavitt/City College of New York, *Structures of Coastal Resilience*, 2015

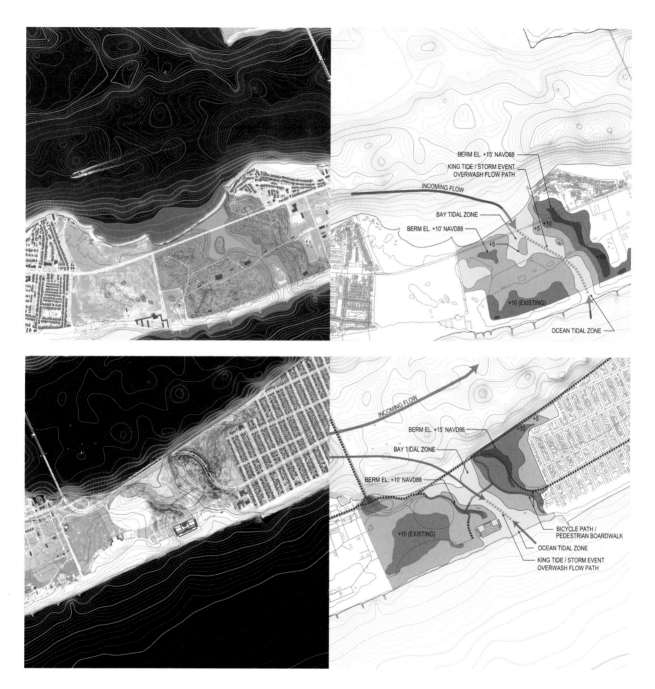

Labels within the figure (top right panel):
BERM EL. +15' NAVD88
KING TIDE / STORM EVENT
OVERWASH FLOW PATH
INCOMING FLOW
BAY TIDAL ZONE
BERM EL. +10' NAVD88
+5
+10
+15
+10 (EXISTING)
OCEAN TIDAL ZONE

Labels within the figure (bottom right panel):
INCOMING FLOW
BERM EL. +15' NAVD88
BAY TIDAL ZONE
+5
+10
BERM EL. +10' NAVD88
+10 (EXISTING)
BICYCLE PATH /
PEDESTRIAN BOARDWALK
OCEAN TIDAL ZONE
KING TIDE / STORM EVENT
OVERWASH FLOW PATH

Plate 10: Proposed overwash plains at Fort Tilden and Jacob Riis Park at Jamaica Bay's Rockaway Peninsula. By allowing surge waters to flow into the bay at low elevation points and then back to the ocean when tides recede, overwash plains both reduce flooding of the more developed portions of the peninsula and back bay and improve the water quality of the bay through increased ocean-to-bay exchange.

Catherine Seavitt/City College of New York, *Structures of Coastal Resilience*, 2015

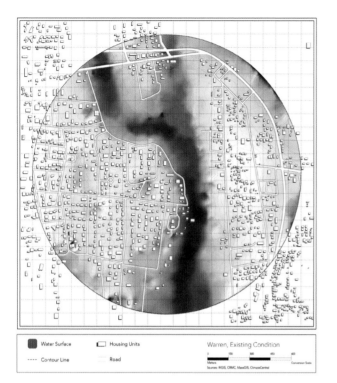

Water Surface Housing Units Warren, Existing Condition

---- Contour Line Road

Meters

0 150 300 450 600 Conversion Scale

Sources: RIGIS, CRMC, MassGIS, ClimateCentral

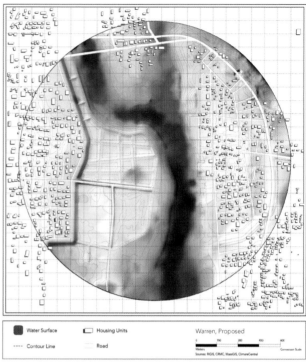

Water Surface Housing Units Warren, Proposed

---- Contour Line Road

Meters

0 150 300 450 600 Conversion Scale

Sources: RIGIS, CRMC, MassGIS, ClimateCentral

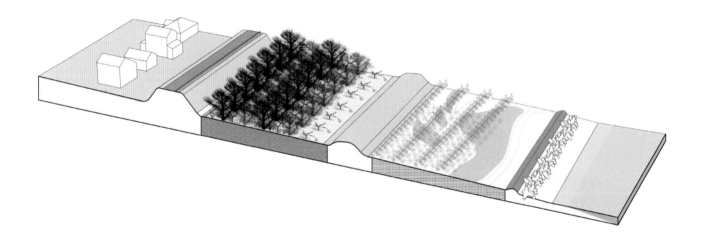

Plate 11: Warren, Rhode Island, with housing and infrastructure currently located along vulnerable low-lying terrain, prone to seasonal flooding and surge events (left). Resettlement to high ground provides an opportunity to create a critical setback and cultivate public parkland (right). The axonometric transect (bottom) demonstrates the layered approach at the littoral edge: settlement, elevated roadway berm, designed forests, thickets, and floodable coastline.

Michael Van Valkenburgh and Rosetta Elkin/Harvard University Graduate School of Design, *Structures of Coastal Resilience*, 2015

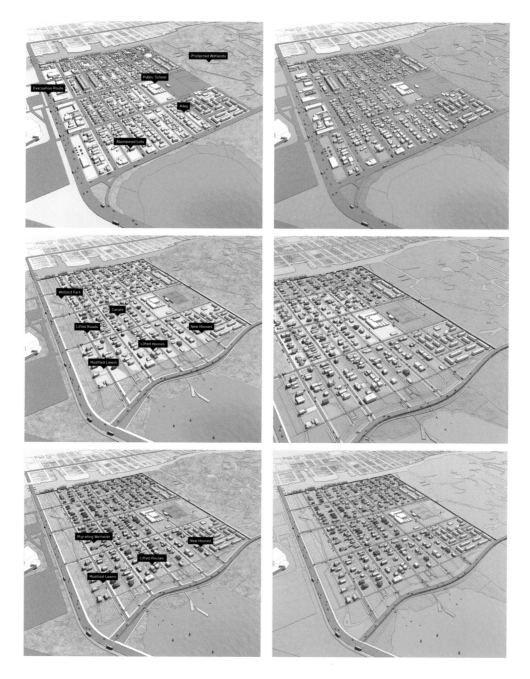

Plate 12: "Amphibious Suburb" proposal for Chelsea Heights, a back-bay neighborhood of Atlantic City and a former salt marsh transformed by suburban development. Phased future development—shown at 2025 (top), 2055 (middle), and 2100 (bottom) in ordinary and surge conditions—would elevate roads and homes, create canals and wetlands, and construct protective edges.

Paul Lewis/Princeton University School of Architecture, *Structures of Coastal Resilience*, 2015

Plate 13: "Water Proving Ground," a proposed crenellated coast at Liberty State Park, Jersey City, producing a landscape of land piers and cross-grains resilient to rising sea levels. The Water Lodge (top) marks the intersection of recreational zones on the land and in the water, and the

Aquaculture Research and Development Center (bottom) provides laboratories and testing beds for estuarine species.

Paul Lewis, Marc Tsurumaki, David J. Lewis/LTL Architects, *MoMA Rising Currents*, 2010

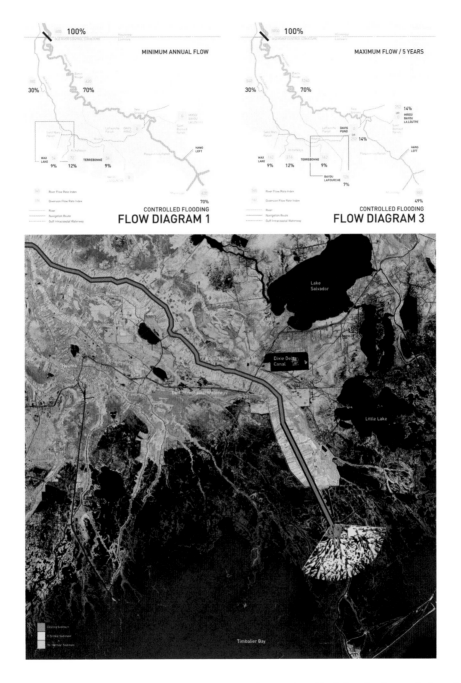

Plate 14: Controlled flooding flow diagrams for the Mississippi River delta at present and in 5 years, with the introduction of diversions to deliver water and sediment to subsiding marshlands throughout the delta. The detailed plan presents the diversion proposal and projected land building for Bayou Lafourche.

Guy Nordenson and Catherine Seavitt/Princeton University with Louisiana State University's Coastal Sustainability Studio, 2010

Plate 15: The SCR storm surge inundation maps present water depth over topography and bathymetry across a gradient scale. The maps are created with probabilistic storm surge rasters that incorporate climate change and sea level rise data projected over the next century.

Catherine Seavitt/City College of New York, *Structures of Coastal Resilience*, 2015

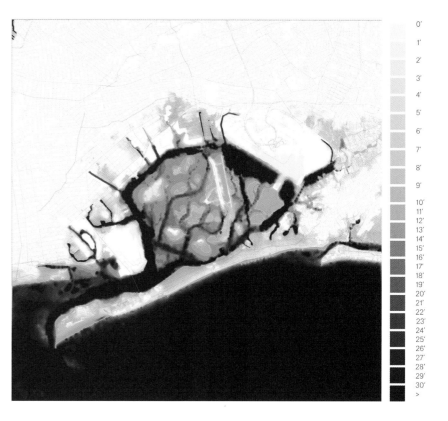

0'	
1'	
2'	
3'	
4'	
5'	
6'	
7'	
8'	
9'	
10'	
11'	
12'	
13'	
14'	
15'	
16'	
17'	
18'	
19'	
20'	
21'	
22'	
23'	
24'	
25'	
26'	
27'	
28'	
29'	
30'	
>	

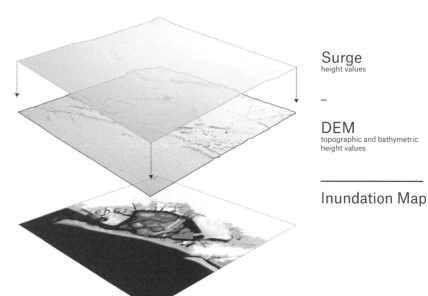

Surge
height values

–

DEM
topographic and bathymetric
height values

Inundation Map

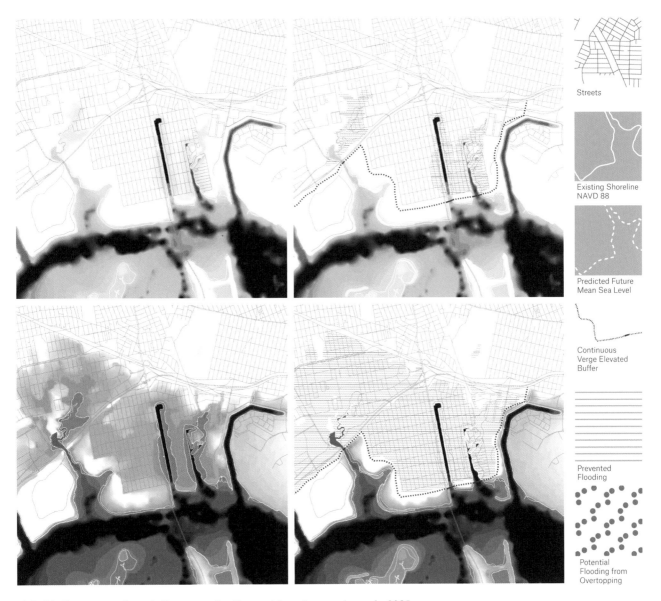

Streets

Existing Shoreline
NAVD 88

Predicted Future
Mean Sea Level

Continuous
Verge Elevated
Buffer

Prevented
Flooding

Potential
Flooding from
Overtopping

Plate 16: Storm surge inundation maps for Howard Beach in Jamaica Bay, showing the effect of topographic modifications on projected probabilistic storm surge inundation for 100-year (top) and 2,500-year (bottom) storms in 2025.

Catherine Seavitt/City College of New York, *Structures of Coastal Resilience*, 2015

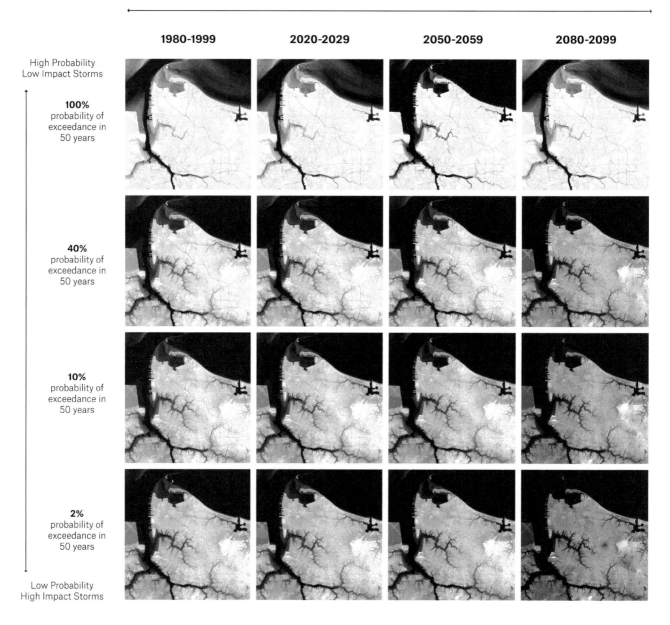

| | 1980-1999 | 2020-2029 | 2050-2059 | 2080-2099 |

High Probability
Low Impact Storms

100%
probability of
exceedance in
50 years

40%
probability of
exceedance in
50 years

10%
probability of
exceedance in
50 years

2%
probability of
exceedance in
50 years

Low Probability
High Impact Storms

Plate 17: SCR map matrix for Norfolk, Virginia, showing projected probabilistic storm surge inundation depths at multiple return periods and over multiple time periods.

Anuradha Mathur and Dilip da Cunha/University of Pennsylvania, *Structures of Coastal Resilience*, 2015

the Empire matches the actual size of the empire. In its ambition of perfection and precision, the map loses all usefulness:

> In that Empire, the Art of Cartography attained such Perfection that the map of a single Province occupied the entirety of a City, and the map of the Empire, the entirety of a Province. In time, those Unconscionable Maps no longer satisfied, and the Cartographers Guilds struck a Map of the Empire whose size was that of the Empire, and which coincided point for point with it. The following Generations, who were not so fond of the Study of Cartography as their Forebears had been, saw that that vast map was Useless, and not without some Pitilessness was it, that they delivered it up to the Inclemencies of Sun and Winters. In the Deserts of the West, still today, there are Tattered Ruins of that Map, inhabited by Animals and Beggars; in all the Land there is no other Relic of the Disciplines of Geography.
>
> —Suárez Miranda,
> *Viajes de varones prudentes*, Libro IV,
> Cap. XLV, Lérida, 1658
> —Jorge Luis Borges,
> translated by Andrew Hurley
> (Collected Fictions, Penguin, 1998)

In order for the world to be navigable, it must be transformed through graphic codes. Carroll's Bellman and crew wander the sea not because oceans have not been charted but because the frame of their map excludes any oceanographic data. Likewise, the "Map of the Empire" lies in "Tattered Ruins" not because it lacked information but rather because the scale of the information presented was too large; the resolution of the data smothered the Empire.

Holiday's "Ocean-Chart" and Borges's Map of the Empire both present a cautionary tale of the role of the navigational chart. Intended as guides for plotting a course, navigational charts make a territory legible to those who move through it. The "Ocean-Chart" and the Map of the Empire, though compelling as images and ideas, fail as tools precisely because they make the world unnavigable. Navigational charts—nautical charts, roadmaps, trail maps, even Google Maps—help the map reader move from one place to another, but their usefulness depends on a static territory. As soon as something shifts—a trail blocked by a fallen tree, a road closed for repairs, a new shipping channel dredged—the chart is outdated and no longer relevant.

Flood Insurance Rate Maps (FIRMs)

Because they are driven by a risk-and-replacement model, the flood maps produced by the NFIP to convey flood risk in coastal and riverine floodplains—FIRMs—function more like navigational charts than like design maps. They help citizens navigate a flood insurance system overlaid on flood-prone territories; however, they do not adequately convey the flood hazards that might occur or indicate how citizens might prepare for them. However, an understanding of the NFIP and FIRMs is important in any discussion of flood mapping in the United States. Recognition of the program's ambitions and shortcomings can aid in the development of new kinds of maps that would be better suited to design and planning, and indeed to the understanding of risk itself.

A history of the NFIP reveals a paradox. Overseen by the Federal Emergency Management Agency (FEMA), the insurance program was created to reduce risk and discourage construction in floodplains, but in many locations it has increased risk by subsidizing insurance rates for flood-prone property owners. In addition, because the federal government will pay for disaster assistance when a flood event occurs, local governments are not always incentivized to plan for flood risk. Administered by FEMA, the NFIP designates flood zones and requires property owners within those zones to purchase flood insurance from the federal government. Private insurance companies tend not to provide flood insurance because of the challenges of creating a rate structure that both reflects a property's risk of flooding and guarantees the company a profit. Flooding is also localized and catastrophic; a single flood could simultaneously damage hundreds or thousands of properties, depleting the insurance company's resources.

Launched in 1968 to provide government-backed insurance for property owners in flood-prone zones, the NFIP uses FIRMs to illustrate three types of hazard areas, each of which correlates to a different insurance rate: special flood hazard areas (SFHAs) represent land that will be inundated in the event of a 100-year flood, or a storm that has a 1 percent chance of occurring per year; moderate flood hazard areas are subject to a 0.2 percent chance storm per year, or a 500-year flood; and minimum or undetermined flood hazard areas exhibit a small yet possible chance of flood. (See Figure 4.3.) The relationship between the FEMA-designated base flood elevation (BFE)—the predicted elevation of water in a 100-year flood—and the elevation at which a structure stands determines its flood insurance rate.[5] The 100-year flood

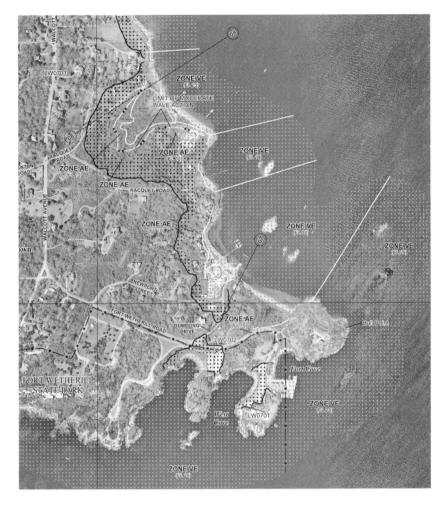

Figure 4.3: Detail of a flood insurance rate map (FIRM) for Newport County, Rhode Island, Panel 176. Map revised September 4, 2013.

National Flood Insurance Program/ Federal Emergency Management Agency

was established as a standard for flood risk assessment in a 1967 report by the Water Resources Council, an important document that offered one of the first sets of "methodologies and standards to be used in developing information about flood hazards."[6] These federal flood maps are a crucial tool for developers, property owners, and citizens; they are used to determine where and how structures are built and the rates at which they are insured.

However, the NFIP fell short of the ambitions of mid-century floodplain management experts. Geographer Gilbert White, who chaired the Task Force on Federal Flood Policy leading to the NFIP, was concerned not only with developing a functional insurance program but with creating an ecologically sound system of floodplain management. According to biographer Robert Hinshaw, White expressed concern that the public emphasis on federally subsidized flood insurance would "divert attention from the broader goal of the task force recommendations: a 'unified national program' for managing not only flood loss/flood control but also floodplains as ecosystems."[7] Over the past half-century, White's concerns have been realized. The U.S. Geological Survey (USGS) and USACE began an extensive hazard mapping initiative in 1966, mapping the extents of all of the 100-year floodplains in the United States, but this effort was driven less by a desire to understand the physical properties and ecological ramifications of potential floods than by a desire to assess financial risk.[8] The program sought to limit development in floodplains by tying higher rates of insurance to properties with a greater probability of incurring flood damage. However, because coastal populations have increased nearly 40 percent since 1970, the policy has not been successful in curbing settlement in flood-prone regions. In fact, subsidized flood insurance premiums at below-market rates provided by Congress to low-income homeowners living in floodplains have encouraged owners to remain in these at-risk locations, while taxpayers bear the cost of repetitive flood loss claims well over the actual value of the homes.[9]

The limitations of FIRMs in planning and design applications stem from their role as insurance instruments. Because they are intended simply to communicate insurance rates to property owners, the maps obscure additional information that is useful to planners and designers. The boundaries of FIRMs are simplified to offer property owners and insurance companies clearly defined zones that correspond to a standardized set of insurance rates; they do not identify the full extent of the floodplain or where flooding may occur. Worse, residual risk—that of the assumed hazards being greatly exceeded, of future sea level rise and other impacts of climate change, or of levees or other flood control structures unexpectedly failing—is unaccounted for in the FIRMs. The lowest-elevation floodplains often are the locations of poorer communities with the least access to resources, creating a situation that is a problem both for science and policy and for social justice. In addition, insurance rates are often subsidized for the poorest communities, encouraging entrenchment in the most dangerous flood-prone territories, and the consequences of risk are apportioned unequally.

The FIRMs conceal detailed hazard information that might help designers and planners create stronger, more resilient, and more equitable coastal communities. For example, although topographic data are incorporated in the process of designating flood zones, this elevation information is not represented in the final maps.[10] The maps also contain a misleading degree of specificity; FIRMs incorporate assumptions regarding topography, bathymetry, tides, storm surge, and wind speed, but these uncertainties are not registered in the final maps. Hard lines designate flood zones, suggesting that buildings within the zone are in danger and those outside are safe, but in reality storm flooding is highly indeterminate. The seemingly static conditions portrayed on the maps are problematic, as noted by National Oceanic and Atmospheric Administration (NOAA) coastal hazard specialist Maria Honeycutt: "Storms are all different. No one storm is going to produce the one-percent flood-risk area. The area that will get flooded is different every time. . . . Floodplains change; the map is not static."[11] FIRMs also rely heavily on historical data. To remain relevant, they require frequent revisions that are often slowly implemented. Property boundaries shift, ground elevations change, adjacent properties are developed, and modifications are made to flood control projects such as levees or reservoirs.[12] Additionally, the synthetic storms created to project storm surge flood inundation for the FIRMs are based on historical precedents and do not incorporate sea level rise or the potential for more intense storms given climate change. The relationship between flood information and property values further complicates mapping efforts. Homes located within floodplains have lower market values than equivalent homes outside the boundaries. Some residents within the floodplain are also required to purchase more expensive insurance. These factors can contribute to political pressure to stall the creation of new maps or refute maps, reducing their accuracy.[13] Designing a project in a coastal floodplain with merely a FIRM for reference is much like setting sail with only the Bellman's "Ocean-Chart."

The deficiencies of these flood maps have real consequences, affecting entire neighborhoods adversely. Falsely assured by FIRMs that their property would remain dry in the face of rising floodwaters and an incoming storm surge, thousands of New Yorkers remained in their homes as Hurricane Sandy made landfall on October 29, 2012.[14] By the time the storm abated, flooded regions in Brooklyn and Queens had greatly exceeded the high-risk areas designated by official maps, and waters rose far above the flood elevations delineated on maps of the affected areas.[15] After Sandy, FEMA updated their maps for this region with greater floodplain extents, an important short-term step for developing new insurance rates. The outcomes of Hurricane Sandy's surge flooding and its powerful predecessor, Hurricane

Katrina—as well as the more recent massive flooding produced by Hurricane Harvey's heavy rainfall—all demonstrate the imperative for an alternative to the FEMA maps and the implementation of a planning-for-hazards approach to development and urban design.

FEMA leaders recognize the shortcomings of the existing flood maps and have taken steps to improve them. Even before Katrina and Sandy, the agency sought to improve the flood mapping system. Funded from 2003 to 2008, the Flood Map Modernization (Map Mod) initiative updated and digitized FIRMs. The effort sought to make maps more dynamic by incorporating environmental changes with greater frequency and accuracy, updating maps with data made possible by new technologies, and encouraging responsible floodplain management through an improved public awareness of flood hazards.[16] Although the project was granted more than $1 billion in federal funding, its scale was immense, and it remained underfunded. Moreover, Map Mod did not explicitly address sea level rise.[17] Commissioned by FEMA and NOAA, a 2009 National Research Council report titled *Mapping the Zone* provided detailed recommendations for improving the accuracy and clarity of flood maps. More nuanced maps might include metrics such as water depth and velocity to show a range of conditions.[18] FEMA continues to update its maps, and some FEMA regions have made significant progress in issuing new editions; Region II, which includes New York and New Jersey, conducted an extensive coastal flood mapping study after Sandy.[19]

The demand for more comprehensive flood maps drives continued efforts to incorporate climate change into coastal hazard mapping. The 2012 Biggert–Waters Flood Insurance Reform Act requires FEMA to consider the "best available science regarding future changes in sea levels, precipitation, and intensity of hurricanes" in its flood maps.[20] In June 2013, NOAA, FEMA, and the USACE collaborated on an interactive tool that would help in the planning and rebuilding of regions affected by Sandy. These maps for New York and New Jersey were not intended to drive insurance ratings or official flood zone designations but rather to help planners and the public understand how elevated sea levels might expand flood prone areas. The maps couple the flood risks of a 100-year storm (1 percent annual chance of flooding as defined by FEMA's flood hazard data) with projections for four sea level rise scenarios for 2050 and 2100.[21] Maps created for the New York City Panel on Climate Change (NPCC), organized by the New York City Mayor's Office of Long-Term Planning and Sustainability, similarly combine the existing FEMA 100-year and 500-year flood zones with sea level rise projections.

(See Figure 4.4.) These efforts are necessary to convey a general sense of how sea level rise will increase coastal flooding, but they do not yet offer comprehensive probabilistic assessments of coastal flood hazards.

The current use of FIRMs in design for coastal resilience points to a dearth of appropriate tools for a planning-for-hazards approach to resilience and adaptation. Coastal planning and design demands *hazard* mapping—mapping that illustrates potential flooding and makes sea level rise and storm surge data legible to communities and planners. Moreover, hazard maps can evaluate the effectiveness of coastal resilience projects in relation to state-of-the-art climate science.

Dynamic Hazard Mapping

Complex variables, shifting in both time and space, make storm surge a particularly challenging hazard to map. Yet mapping possible flooding from storm surge—where it may occur and the heights it might reach—is critical to the planning and development of new designs for coastal resilience. Following a planning-for-hazards rather than a risk-and-replacement approach, design maps for coastal resilience must present sea level rise and storm surge as independent of the financial risk those hazards might create. Hazard maps do not consider the financial value of property when presenting flood-prone areas. They should also incorporate change over time—especially changes driven by climate change but also changes such as erosion and sedimentation that are driven by development or change in use. In short, in order for storm surge maps to be useful as tools for design, they should do three critical things: illustrate hazards spatially, incorporate the changing effects of climate change on those hazards, and allow potential design interventions to be tested against possible hazard scenarios.

Many factors affect storm surge and its subsequent effect on coastal communities. The forms of bathymetry and topography can allow surge to spread over vast territories in shallow depths or funnel into small bays or inlets. The speed, direction, and pressure field of winds within a hurricane determine, to a large degree, the height of the surge and the extents to which it will travel. Surge is also affected by the size of the storm's eye, the path of the storm in relation to the coastline, and whether the storm travels parallel to the coast or approaches it obliquely or directly.[22] Astronomical tides also contribute to surge elevations; if a storm hits at high tide in an area with a large tidal range, flooding may be greater than if the storm had hit at low tide. In sum, the

Figure 4.4: New York City Panel on Climate Change (NPCC) storm surge inundation data at Jamaica Bay for a 100-year return period. FEMA's 2013 post-Sandy preliminary work map (PWM) indicating their assessment of the 100-year return period is indicated in black. Additional inundation due to sea level rise is shown in medium gray (2020s) and light gray (2050s).

Catherine Seavitt/City College of New York, *Structures of Coastal Resilience*, 2015/Data source: NPCC and FEMA

combination of site-specific factors (topography, bathymetry, coastal protection, vegetation) and storm-specific factors (size, shape, speed, path, timing) make storm surge heights and extents difficult to predict.

The effects of climate change add variables to storm surge in the present and future. Rising ocean temperatures, melting ice sheets, and other factors elevate mean tide levels, increasing surge heights by raising the base line from which they are measured. In addition, warmer global temperatures, particularly sea surface temperatures, affect tropical cyclone climatology. Research by Kerry Emanuel and others indicates that warmer ocean temperatures might increase the intensity of hurricanes—future storms may have stronger winds and larger storm surges than storms in the current climate.[23] There is consensus among scientists in the field that although tropical storms may become more intense, their overall frequency may decline. However, the most severe or intense storms may occur with *greater* frequency. In other words, the most damaging storms will be not only more intense but also more frequent.[24]

These factors present challenges, but they also reaffirm the need for flood maps that account for these variables. The interdisciplinary design project SCR sought to incorporate variables into a series of flood maps developed from 2013 through 2015, for four sites along the North Atlantic coast that correspond to the design sites described in Chapter 3. The SCR maps were carefully designed to spatially represent probabilistic coastal flood hazard risk, accounting for climate change over the next century. They also evaluate the flood mitigation effects of design proposals, effectively testing the efficacy of specific interventions that might reduce inundation in possible storm scenarios. This section outlines the mapmaking process conducted for SCR, a methodology that may be applied by other interdisciplinary teams to other projects and sites.

The collaborative mapmaking effort involved not only the design teams developing projects for the four North Atlantic coastal embayment sites but also a mapping team led by engineers Guy Nordenson and Michael Tantala, with a scientific research team at Princeton led by professor of geosciences and international affairs Michael Oppenheimer and professor of civil and environmental engineering Ning Lin. Lin and Oppenheimer, working with Talea Mayo and Christopher Little, focused their coastal flood hazard assessment on the four SCR design sites of Narragansett Bay, Jamaica Bay, Atlantic City, and Norfolk so that their research on sea level rise and surge hazards might accomplish two broad goals. First, the research would provide valuable information for the design process, informing the design teams at the local scale and scope of possible flooding from storm surge in the next century at specific sites, thus influencing the shape and form of the design projects. Second, it would allow the design teams to test their topographic interventions against surge hazard data projected across the twenty-first century. By using these data to create dynamic flood maps—first with the existing topography of a site and then with a modified topography reflective of their design proposals—the design teams could evaluate the effectiveness of their proposed interventions. The vulnerability of communities could be reduced by either mitigating flooding or, more radically, restructuring the urban landscape so that neighborhoods and critical infrastructure are protected.

The SCR flood maps are *hazard* maps, not risk maps. They show inundation from surge, tide, and sea level rise but not the financial or other consequences of that inundation. Intended for design and planning projects at the scale of a neighborhood or city as opposed to a single lot or structure, the SCR maps illustrate not only possible flood scenarios but also the uncertainty associated with indeterminate events in a changing climate. They provide a broad picture of both moderate and extreme coastal flood events and how those events might change over the course of the twenty-first century. (See Color Plate 15.)

Making Flood Hazard Maps

Productive, useful, and successful design maps depend on a thorough analysis of storm surge hazards and how that surge will affect the coast. Flood depths and the impacts of flooding on communities are determined not only by how high a storm surge might rise but also by the shape of the land surrounding the coast, both above and below the water. Hydrodynamics—how topography and bathymetry exert forces on water—shapes how surge makes its way onto otherwise dry land, causing flooding. For any site, a rigorous coastal flood hazard mapmaking process is divided into two critical components: compiling topobathy data (combined topography and bathymetry) and surge modeling. Both topobathy and storm surge are dynamic and malleable; they will change over time with climate change and sea level rise. Topobathy and probabilistic storm surge can be mapped spatially in geographically referenced software, geographic information systems (GIS), and ultimately overlaid to elucidate the interaction between coastal landforms and surge hazards and the flooding that this interaction will produce. In short, flood maps depend on two layers of digital information: topobathy digital elevation models and storm surge raster files. (See Color Plate 15.)

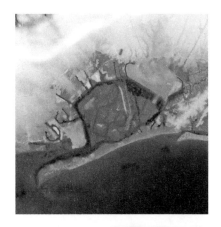

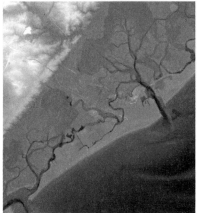

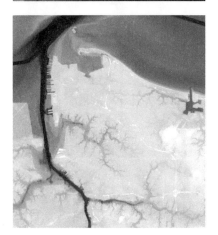

A topobathy digital elevation model (DEM) combines topography (elevational information for land surface) and bathymetry (elevational information about the earth's surface under the ocean) into one continuous digital file, providing an essential tool for understanding the impact of sea level rise and storm surge on a specific region. (See Figure 4.5.) In the digital raster file, each cell within a DEM is assigned a numerical value corresponding to an elevation above or below a designated zero point. Each raster cell might represent a spatial area 3 feet wide by 3 feet long, or 10 feet by 10 feet, or even larger, depending on the resolution of the file. Making a DEM is complex, as it entails compiling data from multiple sources. These data do not always mesh together seamlessly, but compiling them correctly is an important first step to developing a useful flood map based on an accurate representation of the coast.

Topographic and bathymetric data are collected and distributed separately by distinct government agencies. The USGS distributes topographic data, and NOAA provides bathymetric data for bays and coastal waters along U.S. coasts. Topography and bathymetry can also be sourced from FEMA, the USACE, the National Park Service, and state and local GIS information clearinghouses. The SCR design teams sourced digital elevation data primarily from USGS, NOAA, and FEMA and then combined and updated these data sets to create smooth, continuous, and accurate DEMs. (See Figure 4.6.)

With sea level rise and shifting, unstable coastlines, the designation of a reference baseline or datum, the zero-foot elevation, is particularly important for the construction of an accurate topobathy DEM. Coastlines shift constantly, moving from high tide to low tide over a period of hours but also migrating higher over years and decades as sea levels rise. The selection of a datum is essential to tracking that rise, even if the elevation initially set as zero no longer remains the mean tide level over time. The SCR maps use the North American Vertical Datum of 1988 (NAVD 88) as a zero-foot elevation marker. NAVD 88 is the standard orthometric surveying height in the United States, and it falls very close to mean tide level at the four SCR sites. (See Figure 4.7.) Some topographic and bathymetric data sets use other local datums as the zero-foot elevation, and the selection of a datum should be made for each site based on the data sets available. NOAA bathymetric navigational charts often set water depths relative to a zero set at the local mean low tide level. To accurately merge topographic and bathymetric datasets, datums must be adjusted so that the topography begins where bathymetry ends, creating a seamless transition. With sea level rise, tidal zones may shift, and high, mean, and low tide datums may move accordingly. Standardized

Figure 4.5: Digital elevation models of Jamaica Bay, New York; Atlantic City, New Jersey; and Norfolk, Virginia.

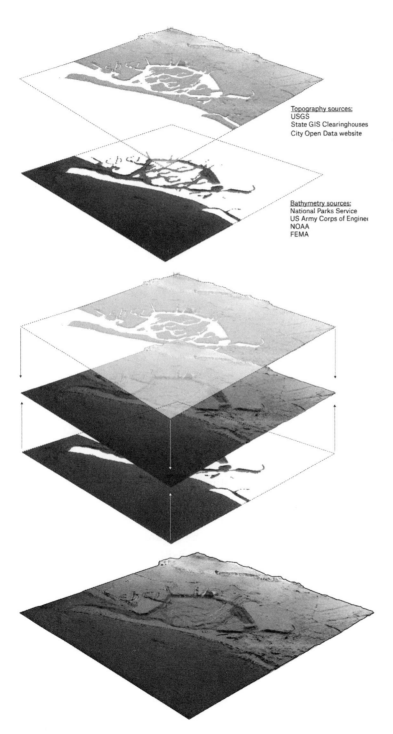

Topography sources:
USGS
State GIS Clearinghouses
City Open Data website

Bathymetry sources:
National Parks Service
US Army Corps of Enginee
NOAA
FEMA

Figure 4.6: Digital elevation models are developed from topographic and bathymetric data, often sourced separately. An integrated topobathy model, a continuous surface that traverses land and water, allows comprehensive digital modeling of storm surge inundation at the coast.

Julia Chapman, *Structures of Coastal Resilience*, 2015

Figure 4.7: The process of merging topographic and bathymetric data requires close attention to vertical datums. The SCR surge maps use NAVD 88 as the vertical datum for referencing 0' elevation.

Julia Chapman, *Structures of Coastal Resilience*, 2015

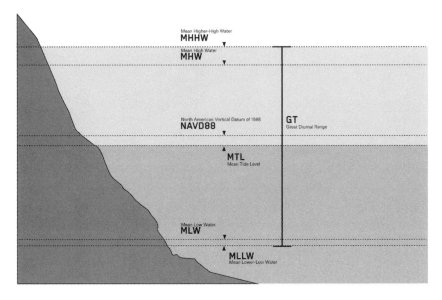

vertical datums such as NAVD 88 become increasingly important reference elevations in the process of measuring the change of water levels and surges over time.

Topobathy DEMs create accurate digital representations of the surface of the earth in its present state, but DEMs can also be manipulated to reflect proposed changes to topography and bathymetry. Each SCR design team created a base DEM to represent the earth's surface in the present. The design teams then modeled the topographic modifications—berms, verge enhancements, atoll terraces, "fingers of high ground," canals, and channels—in their design proposals to determine the performance of particular interventions under varied surge scenarios in both the present and future climate. To adjust a DEM to incorporate design interventions, new contours or elevation points from a digital drawing file are imported into the GIS map of the existing conditions. A new, modified DEM might include proposed berms, elevated roads, canals, expanded marshland, raised landforms, or new lowered grounds for retaining floodwaters. Recognizing that large-scale territorial transformation takes significant time—even decades—the SCR design teams ultimately developed a series of DEMs representing the shape of their modified sites at distinct intervals through the twenty-first century, for the years 2025, 2055, and 2090. The phased design process represented through these DEMs also reflects the increasing sea level rise and surge levels, given climate change, projected across the century.

If topobathy DEMs are the base layer, or surface, of the dynamic flood mapping process, surge raster files are the secondary layer, or fill, moving on top of this surface. The continuous topobathy can be thought of as an empty vessel and the surge raster as the liquid poured into that vessel. To make surge data raster files, surge elevation values are plotted onto a grid and then interpolated digitally into a continuous three-dimensional surface, not unlike a DEM. The floodwaters of a surge do not just fill the areas of lowest elevation to find a horizontal level; they also have their own three-dimensional shape, rising higher in certain areas in response to wind and hydrodynamic forces. In digital form, the elevational data of surge rasters is informed by the shape of the land underneath but also by other environmental and meteorological factors. Surge raster files, though plotted digitally in ArcGIS, depend on data points produced with climate models and hydrodynamic models.

A storm surge raster file articulates not the physical shape and surface of the ocean at a particular moment in time but rather the shape of a *probabilistic* still water surge elevation for a particular return period, incorporating sea level rise and tide data, also determined probabilistically. (See Figure 4.8.) In other words, each spatially located data point within the surge raster surface is created by combining the mathematical *chance* that surge will reach a certain height based on information about possible storms and when those storms might occur in the tidal cycle. Because so many factors determine surge height—wind, hydrodynamics, waves, seawalls, jetties, piers, vegetation—some choices may need to be made regarding variables to include and variables to leave out or incorporate in another way. For example, the SCR surge raster data points do not include the additional height that could be reached by individual waves; these data were excluded given that the sites of the SCR designs were developed at coastal embayments, where the impact of waves is generally minimal.

To develop a final flood map representing flood inundation depths, a DEM must be numerically subtracted from the surge raster surface in GIS. For each raster cell, the software numerically subtracts the topobathy (land) elevation from the surge (water) elevation. The result is a digital raster file that shows the depth of the total water column—the vertical extent of water above the elevation of topography or bathymetry. (See Figure 4.9.) This method visualizes the surge elevation on land as well as over water, illustrating the impact of storms on offshore vegetation or structures. This presentation method also blurs the boundary between land and water—an appropriate representation in areas where tidal zones may shift over time with sea level rise. (See Color Plates 15 and 16 and Figure 4.10.)

Figure 4.8: Surge raster files are produced from a series of data points that represent the surface of a flood. Projected flood elevation data are processed for the center point of each cell of the Sea, Lake, and Overland Surges from Hurricanes (SLOSH) basin that overlaps the study site.

Julia Chapman, *Structures of Coastal Resilience*, 2015

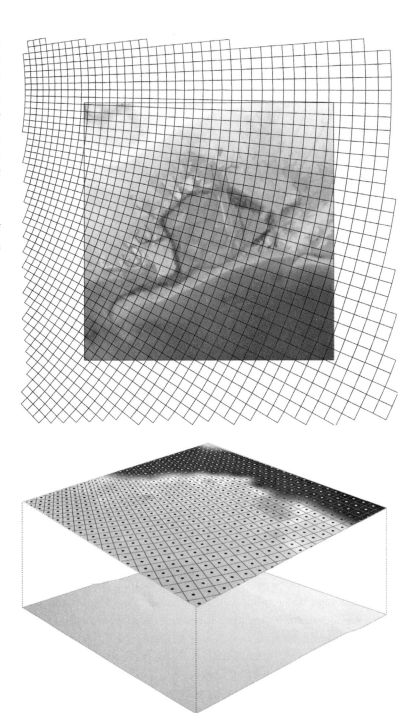

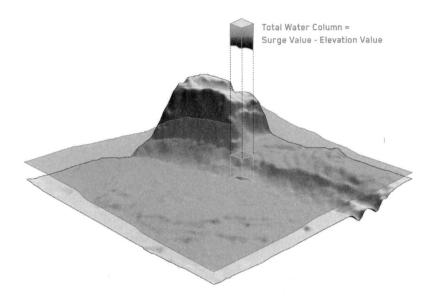

Total Water Column =
Surge Value - Elevation Value

Figure 4.9: The SCR surge inundation maps chart the depth of water above an underlying topobathy surface.

Julia Chapman, *Structures of Coastal Resilience*, 2015

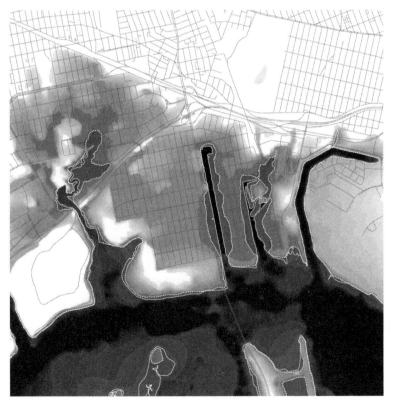

Figure 4.10: Storm surge inundation map of Howard Beach in Jamaica Bay, showing the impact of a projected storm event for 2050 with a 500-year return period.

Catherine Seavitt/City College of New York, *Structures of Coastal Resilience*, 2015

Climate change is complex, and its impacts are indeterminate. Current tools and data analysis do not yield certain answers. Any mapmaking process must reconcile with the shortcomings of available data sets while acknowledging and accounting for data that are available. The limitations of the SCR maps include the fact that the probabilistic surge hazard data do not account for the duration of flooding. Because surge elevation data are neither time-based nor associated with a particular storm event, they do not indicate how long a surge may last. The maps therefore do not illustrate the extent of flooding from overtopping berms in extreme storm events. To account for these limitations, the SCR maps use graphic hatches to represent areas where overtopping might occur and areas where topographic intervention has prevented flooding. (See Color Plate 16.) When a proposed design moves toward implementation, more detailed hydrodynamic analysis can be used to determine this type of information. Unpredictable damage from surge is also not easily incorporated into flood maps. SCR maps do not consider the consequences of surge hazards. For example, if a surge causes a levee or seawall to fail, uplift, or rupture, the maps will not illustrate the extent of flooding that results from that failure. Likewise, if a surge causes a permanent breach in a barrier island, the maps will not show the additional flooding spurred by the breach. The SCR maps serve the purpose of prioritizing large-scale change over time above a detailed analysis of local impacts. Their purpose is to view flood impacts holistically and regionally and to support design strategies for resilience that operate at the scale of that region over time, given the future impacts of sea level rise and storm surge.

Map Matrices

The range of possible flood scenarios produced by the SCR hazard mapping method is central to its potential as a design tool. Rather than create a single definitive flood map, the method supports a matrix of possible maps, enabling planners and designers to see opportunities for design intervention across a spectrum of flood conditions over time, at multiple return periods, and with respect to inevitable climate change. Working with this multiplicity of conditions, designers might be empowered to create landscapes and buildings that do not resist floods but rather respond to them. Though used here for four select sites, the mapping method and matrix framework could be transferred to many other contexts. At larger scales, it could function as a design and planning tool for coastal communities and serve as a model for a holistic probabilistic coastal flood hazard mapping endeavor at the scale of a region or even the entire United States.

The matrix presents flood inundation maps derived from observed tropical storm scenarios for four time periods: 1980–1999, 2020–2029, 2050–2059, and 2080–2099. For each period, probabilistic flooding is shown for the 1-year, 100-year, 500-year, and 2,500-year storm events, which correlate to 100 percent, 40 percent, 10 percent, and 2 percent probability of exceedance in 50 years. The SCR science research team assessed storm surge hazards for the end of the twentieth century, based on observed storms that occurred between 1980 and 1999, and the end of the twenty-first century, based on synthetic storms described in the next section of this chapter. From these two sets of data, they then interpolated surge hazards for the 2020s and 2050s. To assess surge at these intermediary time periods, calculations were made to determine how the probabilistic return periods might shift as the century progresses. For example, the surge elevation of the 100-year return period at the end of the twentieth century may become that of a lower return period—perhaps the 50-year return period—at the end of the twenty-first century. As high surge elevations become more frequent over time, extreme surge events also yield more extensive flooding.

The SCR site-specific map matrices not only quantify hazards but also provide a means to evaluate possible design interventions intended to mitigate those hazards. Each matrix registers both existing conditions and proposed topographic changes and structural interventions to demonstrate the effects of a range of storm events coupled with projected sea level rise. For each site, the design and mapping teams produced thirty-two distinct maps to present a broad spectrum of surge scenarios with varying levels of likelihood at distinct intervals throughout the twenty-first century. (See Figure 4.11 and Color Plate 17.) The first matrix of sixteen maps merges the surge rasters with the current, present-day DEMs to show how these surge scenarios would affect regions if no action were taken and the terrain left unmodified. The second matrix of sixteen maps merges the surge rasters with the modified DEMs with future phased topographic transformations to the terrain. These maps therefore show the effect of proposed topographic modifications on flood inundation depths, even as surge levels increase over the course of the century. They also illustrate how some low-probability events are so extreme that even dramatic topographic change will have little impact in reducing inundation.

Figure 4.11: Map matrix for Norfolk, Virginia, showing probabilistic storm surge inundation over the next century for multiple return periods. Flood conditions for the existing terrain (top) and the modified terrain, including the proposed "fingers of high ground" (bottom), allow comparison.

Anuradha Mathur and Dilip da Cunha/University of Pennsylvania, *Structures of Coastal Resilience*, 2015

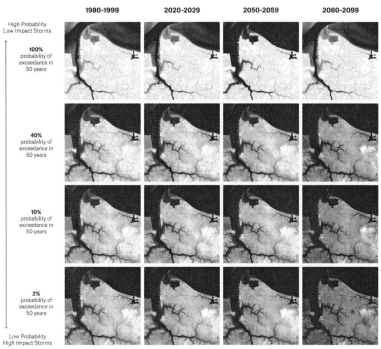

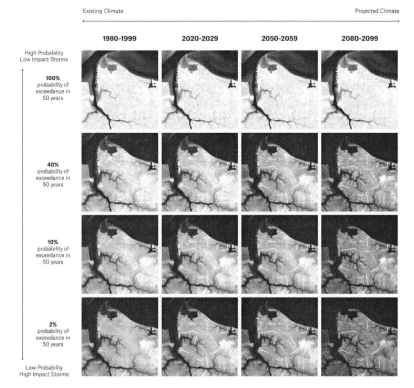

When viewed as a whole, the matrix both reveals the increased chance of dramatic surge hazards over time in a specific location and the mitigation of those hazards through progressive intervention. For example, the matrix of Anuradha Mathur and Dilip da Cunha's proposal for Norfolk (described in Chapter 3) shows that the elevation and expansion of an existing spine of raised land at Lambert's Point can, over time, create a viable and habitable urban strip despite a rapidly increasing risk of deep inundation in the lower surrounding areas. (See Figure 4.12.)

The Norfolk map matrix region illustrates the dramatic possibility of complete inundation throughout the region with a severe event, but the Narragansett Bay map matrix presents a different story. (See Figure 4.13.) In Rhode Island, dramatic elevational shifts between low-lying areas—coastal reservoirs, river deltas, and island valleys—and much higher uplands result in the surge inundation projected for the end of the century with low-probability storm events being equivalent to that projected to occur much sooner with higher-probability storm events. The Narragansett Bay designers, Michael Van Valkenburgh and Rosetta Elkin, propose topographic interventions at a localized scale in areas with low elevation. Their map matrix for the Hummocks reveals that a minor intervention—a berm alongside an evacuation route—can protect vital infrastructure and potentially save lives in a minor or severe storm scenario. (See Figure 4.14.) In Newport, on the southern coast of Aquidneck Island, the designers propose elevating topography in two locations along the coast to protect the city beyond. (See Figure 4.15.) The Rhode Island matrices illustrate many of the arguments for resilient design made in Chapter 3, demonstrating that localized and strategic topographic intervention can have a profound effect in the face of both low- and high-probability storms.

Paul Lewis's SCR project for Atlantic City includes a matrix for both the entire city and the much smaller neighborhood of Chelsea Heights, the location of his design proposal. The first matrix demonstrates that in less severe events, much of the flooding will occur not at the ocean front but along the back side of Absecon Island as surge flows through inlets to the back bay. Dunes and nourished beaches will protect the ocean-facing side of the city. (See Figure 4.16.) In more severe storms as sea level rise increases, the ocean-front dunes will become inadequate for protection, and surge may inundate the full extent of the city. In the back-bay neighborhood of Chelsea Heights, a comparison between the existing topography and proposed topography matrices reveals that the proposed design interventions—perimeter berms, elevated streets, and lowered canals—do not keep the neighborhood fully dry; rather, they protect critical infrastructure and homes while allowing

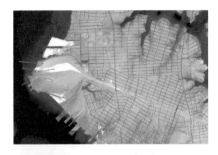

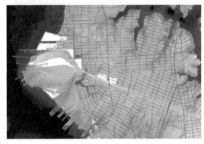

Figure 4.12: Storm surge inundation maps of Lambert's Point, Norfolk, before and after design modifications to the existing railway ridgeline. The maps reflect a projected flood event in the 2020s with a 2,500-year return period.

Anuradha Mathur and Dilip da Cunha/University of Pennsylvania, *Structures of Coastal Resilience*, 2015

Figure 4.13: Map matrix for Narragansett Bay, Rhode Island, showing probabilistic storm surge inundation over the next century for multiple return periods.

Michael Van Valkenburgh and Rosetta Elkin/Harvard University Graduate School of Design, *Structures of Coastal Resilience*, 2015

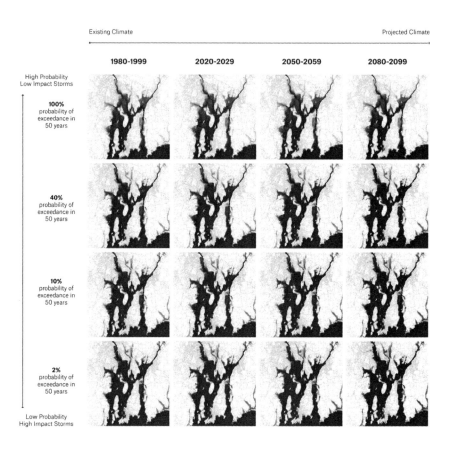

Existing Climate

Projected Climate

| 1980-1999 | 2020-2029 | 2050-2059 | 2080-2099 |

High Probability
Low Impact Storms

100% probability of exceedance in 50 years

40% probability of exceedance in 50 years

10% probability of exceedance in 50 years

2% probability of exceedance in 50 years

Low Probability
High Impact Storms

water to move through and eventually out of the neighborhood after a surge event. (See Figure 4.17.)

Catherine Seavitt's SCR Jamaica Bay map matrices illustrate potential flood inundation at two scales. One matrix of maps includes the full bay and the adjacent neighborhoods of Brooklyn and Queens, conveying the far-reaching impact of the design team's proposed elevation of the verges along the Belt Parkway, which could protect multiple neighborhoods and hundreds of thousands of New Yorkers from flooding. (See Figure 4.18.) A second map matrix, detailing the Howard Beach neighborhood of Queens, shows more clearly how this verge enhancement is strategically placed to connect with existing areas of high elevation, protecting the low-lying points where water would rush inland. (See Figure 4.19.)

Although the SCR map matrices are unique in their precise configuration, other flood map suites offer pertinent precedents for illustrating multiple scenarios and unpredictable events. Several European flood maps demonstrate innovative graphic techniques for relaying information about flood hazards. In accordance with a 2007 directive, each European Union member nation has produced a suite of digital flood hazard and risk maps that assess the hazards and risks of low-, medium-, and high-probability events.[25] In some countries these maps are part of binding regulations for development within flood zones, and in others they remain advisory guidelines. Instead of simply showing the boundaries of flood-prone regions, as FIRMs do, many European flood extent maps are coupled with water depth data to illustrate a gradient of flood severity. Some also incorporate the velocity of moving water; a German hazard map defines the "intensity" of a riverine flood as a combination of water depth and flow velocity.[26] Alternatively, the national flood maps of Belgium and Ireland represent the extents of three different probabilities of flooding in shades of blue.[27] Extent maps might be supplemented with point data or include overlays of urban information to provide context. Denmark's online map viewer provides land use information at flood inundation zones. French flood hazard maps for the Aquitaine region are compiled into a suite; one map synthesizes extensive flood information by illustrating flood extents for rare, medium, and frequent events, and additional maps detail probable flood depths for these events.[28] In some cases, these maps also account for climate change with an additional flooding scenario for "medium" events.

Several recent programs and simulations illustrate predicted flooding from sea level rise, including Surging Seas, a tool developed by the independent organization Climate Central, and the Sea Level Rise and Coastal Flooding Impacts interactive map, developed by NOAA. However, these online programs do not incorporate the increased severity of storm surge that may accompany climate change and rising sea levels. In addition, the interactive tools do not spatially illustrate the likelihood of specific sea level elevations; instead, they allow users to see how a broad range of sea level increases, from 1 to 10 feet, might affect coastal communities.

The SCR maps and matrices are intended as a precedent example for future flood mapping investigations, which may enrich and elaborate the process described here. In 2015, the matrices were uploaded to the SCR website for visitors to use interactively. They also provided the starting point for a conversation on hazard mapping held as a summit of coastal engineers, climate scientists, and geographers that took place at the offices of the Rockefeller Foundation in May 2015. The conference demonstrated the widely held sentiment, from academics and professionals across disciplines, that

Figure 4.14: Storm surge inundation maps of the Hummocks, Narragansett Bay, reflecting a projected flood event in the 2050s with a 100-year return period. The proposed design at this location (bottom) protects a critical evacuation route from inundation.

Michael Van Valkenburgh and Rosetta Elkin/Harvard University Graduate School of Design, *Structures of Coastal Resilience*, 2015

Figure 4.15: Storm surge inundation maps of Sachuest, Rhode Island in the 2020s for a storm event with a 100-year return period. Design interventions at this location (bottom) prevent freshwater reservoirs from flooding and protect coastal neighborhoods and roads.

Michael Van Valkenburgh and Rosetta Elkin/Harvard University Graduate School of Design, *Structures of Coastal Resilience,* 2015

hazard mapping in the United States critically needs academic attention and public investment. The SCR maps were produced within several months and with widely available technology. With a larger investment, design maps that account for a changing climate could be developed nationally as a critical tool for building resilience along the coast. At the same time, smaller-scaled grassroots and academic projects might also build on the techniques described here, producing design maps relevant for specific planning and design initiatives at other coastal sites.

Producing Dynamic Data

Future storms will not be exactly like past storms. Both maps and their underlying data must consider the nature of that change. The current SCR design maps can only present surge hazards probabilistically because they represent data derived probabilistically. These data account for the impact of climate change on sea level but also for the climatology of hurricanes and tropical storms. Because the SCR mapmaking process was interdisciplinary, the data and the maps were produced simultaneously through the close collaboration of science, design, and mapmaking teams. The data were the result of developing research in tropical storm climatology and hydrodynamics but also part of a conversation about the creation of data and maps that would best aid designers in refining and evaluating their proposals. The SCR data assessment described in this section is envisioned, like the mapmaking process, as a prototype for future research and analysis.

Creating probabilistic storm surge data is challenging because hurricanes are rare, and thus scientists face a limited sample size of documented historic events. Since the National Weather Service began recording hurricane data in 1851, 293 hurricanes have made landfall on the U.S. mainland through 2017. This may seem like many storms, many demanding years of recovery efforts, but for scientists who study hurricanes and tropical storms, the number presents a very small sample size from which to extrapolate future events. Each of these 293 hurricane events followed a different track and made landfall at a unique location, affecting coastal communities with different hazards—wind, storm surge, rain, and even snow. The next storm to strike the coast will be different from any previous storm. The history of storms and the vulnerability of locations do not always align; some low-lying coastal communities have been very lucky to escape the path of a serious hurricane and some seemingly protected regions have been hit hard by improbable storms.

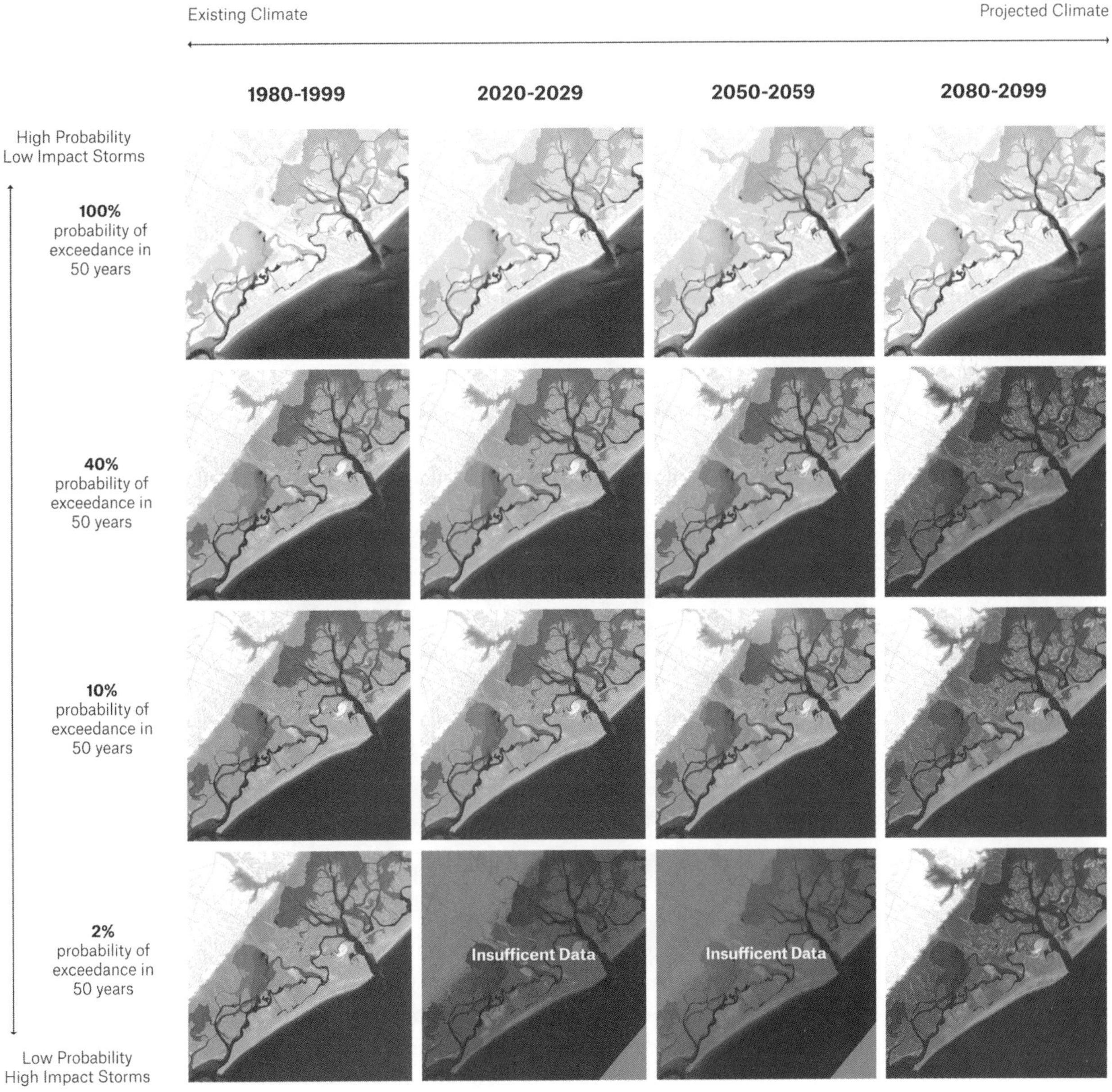

Figure 4.16: Map matrix for Atlantic City, New Jersey, showing probabilistic storm surge inundation over the next century for multiple return periods. Due to insufficient data, the matrix lacks maps for the most extreme storms in the 2020s and 2050s.

Paul Lewis/Princeton University School of Architecture, *Structures of Coastal Resilience*, 2015

Figure 4.17: Storm surge inundation maps of the back-bay neighborhood of Chelsea Heights, Atlantic City in the 2080s, before and after design interventions. The top row illustrates flooding from a 1-year return period; the bottom row represents flooding from a 2,500-year return period. The design proposal does not aim to keep the entire neighborhood dry; rather, it allows water to move through canals and wetlands while elevating homes and roads.

Paul Lewis/Princeton University School of Architecture, *Structures of Coastal Resilience*, 2015

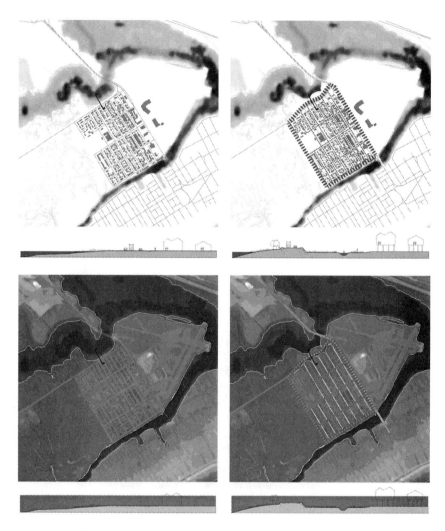

The challenge facing scientists and designers seeking to assess future storm surge hazards is therefore twofold: first, to overcome the limited sample size of historical storm events by simulating a wide range of storms; and second, to incorporate the effects of climate change into the design of those possible storms. Some methods simulate synthetic storms by shifting historical tracks or combining historical storm data, but this technique does not incorporate climate change. The degree to which synthetic storms adhere to historical events also varies; some simulations may simply adjust the path of a historical storm, and others may create entirely new combinations of variables such as path, size, and wind speed. Synthetic storm simulations that

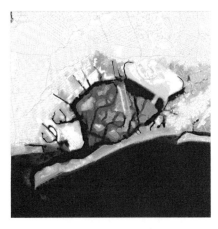
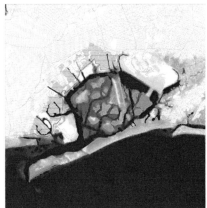
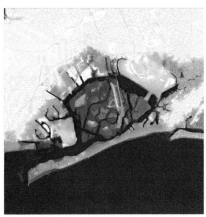
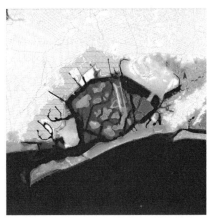

Figure 4.18: Storm surge inundation maps of Jamaica Bay, New York given a 100-year return period in the 2020s (left column) and the 2090s (right column). The top row illustrates existing conditions; the bottom row includes the topographic design interventions of verge enhancement, atoll terracing, and overwash plains.

Catherine Seavitt/City College of New York, *Structures of Coastal Resilience*, 2015

incorporate climate change combine the physical science of hurricanes with future climate data generated from global climate models, computational models that mathematically synthesize atmospheric, oceanic, and land surface processes.[29]

Assessing Hurricane Storm Surge in a Changing Climate

For SCR, a team led by Princeton professors Ning Lin and Michael Oppenheimer with postdoctoral fellows Talea Mayo and Christopher Little assessed coastal flood hazards for four sites along the North Atlantic coast through an innovative method that combines storm surge, sea level rise, and astronomical tide to provide a comprehensive picture of probabilistic flood inundation over the next century. This method builds on techniques developed by

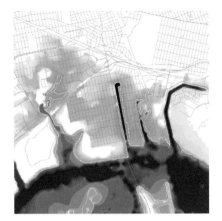

Figure 4.19: Storm surge inundation maps of the neighborhood of Howard Beach, Jamaica Bay, reflecting a projected flood event in the 2020s with a 100-year return period. The proposed topographic verge enhancement and inlet flood gates (bottom) significantly reduce otherwise extensive flooding.

Catherine Seavitt/City College of New York, *Structures of Coastal Resilience*, 2015

Massachusetts Institute of Technology (MIT) scientist Kerry Emanuel along with Lin, Oppenheimer, and others for incorporating inputs from global climate models into synthetic storm simulations.[30] Because it is probabilistic, the method quantifies the extent and depth of inundation at multiple return periods, providing designers with adequate information to plan for multiple levels of risk. Whereas Oppenheimer and Little assess local sea level rise, Lin and Mayo evaluate storm surge hazards and combine those hazards with tide and sea level rise distributions.

Critically, a probabilistic hazard assessment does not evaluate one probable storm but rather many storms, deriving potential outcomes from a series of possible events. Lin models the surge produced by thousands of synthetic tropical storms and hurricanes that pass through particular coastal regions of focus: Norfolk, Virginia; Atlantic City, New Jersey; Jamaica Bay, New York; and Narragansett Bay, Rhode Island. They use synthetic storms simulated by Emanuel's team at MIT through a statistical-deterministic method. The storm tracks are generated statistically and seeded at random origin points corresponding with historical tropical cyclone data, then forced through time and space with randomized climate data. The storm intensity is generated deterministically, meaning that each storm follows basic physical principles. Importantly, only the genesis points are derived from historical storm data; the storm tracks and intensities depend on statistical randomization tied to climate data.[31] Thousands of storms are generated for the entire Atlantic region, and only storms that pass through the specific regions of interest are selected for the assessment.

Driven by the physical science of hurricanes rather than previous storm events, this storm simulation method allows researchers to incorporate inputs from observed historical climate data or global climate models to generate storms that could occur in a past or future climate. Observed historical climate data in this case come from a meteorological reanalysis, produced by the National Centers for Environmental Prediction (NCEP) and the National Center for Atmospheric Research (NCAR), which aggregates and synthesizes global climate data from 1948 through the present. A particular subset of that database can be used to generate storms; in this case, past climate data are limited to the period from 1980 through 1999. To evaluate how storms may change in a future climate, additional synthetic storms are generated with climate data from four global climate models produced by four research centers: the NOAA Geophysical Fluid Dynamics Laboratory (GFDL) in Princeton, New Jersey; Hadley Centre (HadGEM) in Exeter, United Kingdom; the Meteorological Research Institute (MRI) in Japan; and the Max-Planck Institute for Meteorology (MPI) in Germany. These models simulate how the many variables that determine global climate will be affected by greenhouse gases and the continued warming of the atmosphere. Because each model relies on specific assumptions about the interaction of climate variables, simulating storms with multiple global climate models provides a broad picture of possible future scenarios.[32] Each model can also simulate climate change according to multiple greenhouse gas concentration scenarios, estimating future climate in response to variable progress in mitigating carbon emissions. Of the four Representative Concentration Pathways (RCPs)—time-based scenarios of emission increases and reductions over the next two hundred years—included in the Intergovernmental Panel on Climate Change (IPCC) 2014 report, the SCR research team uses RCP8.5, which assumes that global mean temperature and sea level will continue to rise in tandem due to increasing atmospheric concentrations of greenhouses gases through the twenty-first century.[33]

Hydrodynamic Modeling

Storm simulation is the first step in assessing storm surge, and the second step involves understanding how the surge generated by those storms will interact with coastal landforms. This process uses hydrodynamic models that illustrate the movement of wind-driven water in shallow bathymetry, along the coast, through inlets and bays, and on land. Storm simulation and hydrodynamic surge analysis are both complex workflows, carried out by engineering researchers in computational lab settings. Understanding how these analyses work is critical for anyone making flood maps or designing coastal resilience projects. Storm surge data can vary tremendously based on the computational models used. Knowledge of the strengths and weaknesses of the data is important for designers, planners, and community members as they consider how to use and interpret flood maps.

Two hydrodynamic models—ADCIRC and SLOSH—simulate storm surge, and each has particular strengths. The Advanced Circulation Model (ADCIRC) was developed by Rick Leuttich and J. J. Westerink in 1992 for continental shelves, coasts, and estuaries.[34] ADCIRC offers detailed hydrodynamic analysis for coastal regions, but it demands significant computing power. For this reason, ADCIRC is used when an analysis requires a small number of storm simulations. (See Figure 4.20.) Whereas the irregular, flexible mesh grid of ADCIRC allows modelers to control the amount of detail in an analysis, the

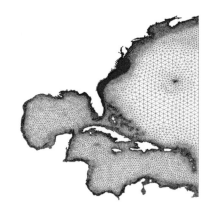
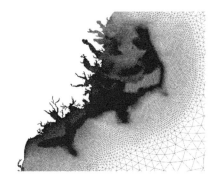
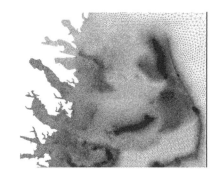

Figure 4.20: ADCIRC grids at the scale of the entire Atlantic and Gulf coasts of the United States (top), the coast of North Carolina (center), and northern North Carolina's Outer Banks (bottom).

Courtesy of Rick Luettich/University of North Carolina at Chapel Hill, Institute of Marine Sciences

Sea, Lake, and Overland Surges from Hurricanes (SLOSH) model operates on fixed polar grid basins. These basins radiate outward from major urban centers, with smaller grid cells and more detailed results around cities and larger grid cells in more remote areas further inland. (See Figure 4.21.) Developed by the National Weather Service, SLOSH is more computationally efficient than ADCIRC. The National Weather Service uses SLOSH to forecast hurricanes in real time as they approach the United States and to determine the vulnerability of coastal areas for evacuation and emergency management. Some research methods run SLOSH for a large number of storms, then identify the most interesting or severe surge events, and finally run those storms with ADCIRC.[35]

Lin and Mayo's analysis of the SCR sites uses SLOSH because of the large number of synthetic storms—3,000 for each site for both the recent and future climates. Their method consists of running each storm through SLOSH to calculate surge inundation at every grid cell, and then plotting these storm events on a histogram. Those events in the "tail" of the histogram—extreme events with high storm surge but a low probability of occurrence—are then inserted into a probability curve using the generalized Pareto distribution (GPD). The GPD creates a continuous curve from a "tail" with a very small sample size—often just a few data points. With probability on the horizontal axis and surge elevation on the vertical axis, this new probability curve allows the determination of surge elevations at specific statistical intervals, for example at the 100-year or 500-year return periods.

Astronomical tide can also have a significant effect on surge elevations. If a storm makes landfall during high tide, surge heights may be far greater than if the storm had arrived 12 hours earlier during low tide. To account for this variability, each synthetic storm is combined probabilistically with the time sequence of the local tidal cycle. This means that the highest surge elevation within each storm is randomly assigned to occur at some point between high and low tide. A broad spectrum of possible tide scenarios is therefore incorporated into the distribution of surge scenarios for a particular location. The combination of tide and surge is called storm tide.

Incorporating Local Sea Level Rise

The computational analysis up to this point accounts for climate change through the use of projected climatology in simulated storms, but predictions for local sea level rise must also be incorporated into the surge hazard data.[36]

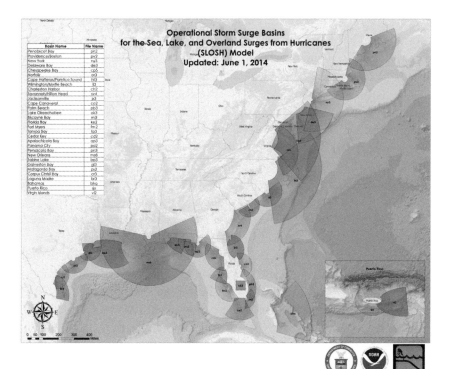

Figure 4.21: Locations of the National Hurricane Center's SLOSH basins for the Atlantic Coast and the Gulf Coast, with an enlarged view of the New Orleans SLOSH basin (below).

NOAA/National Hurricane Center

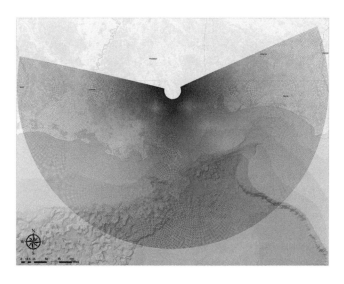

Predominantly determined by the melting of the ice caps and ice sheets, sea level rise will not affect coastal locations uniformly. Factors relating to ocean properties and dynamics, the earth's gravitational field, and nonuniform vertical land motion mean that sea level will rise higher in some parts of the globe.[37] Local sea level rise on the North Atlantic coast of the United States will probably be greater than the global mean. Even along the eastern seaboard, local sea level rise varies. For example, land subsidence in Norfolk, Virginia further elevates mean sea level. Subsidence in this region is caused primarily by the compaction of aquifers due to groundwater extraction and glacial isostatic adjustment—the shifting of the earth as once-glaciated areas rebound from the ice age that occurred twenty thousand years ago.[38]

For SCR, Lin and Mayo combine the storm tide distribution for surge events in the future climate with a local sea level rise distribution to account for the site-specific variation in sea level rise over the next century. To project sea level rise for the four SCR locations, Oppenheimer uses a probabilistic method developed with statistician Bob Kopp and others in 2014. This method relies on climate models, expert assessment, and tide gauge observations of sea level rise to date. The full distribution for each location includes 10,000 potential increases in sea level, demonstrating an extensive range of scenarios for the end of the century. Lin and Mayo combine the storm tide and sea level rise distributions through the combined probability equation, a standard equation for merging two probability distributions. The final output of combined surge, tide, and sea level rise is therefore fully probabilistic. For example, a projection of an event with a 40 percent probability of occurrence at the end of the century includes sea level rise, storm surge elevations, and the chance that the event will occur at high or low tide. This combination of probabilistic distributions accounts for the many factors that will make storm surge in the future distinct from storm surge in the past, including changing storm climatology and rising sea levels, in such a way that considers not a best-case scenario or a worst-case scenario but rather a range of scenarios, each designated by how likely it is to occur. This kind of assessment provides designers, planners, and communities with a full picture of where, when, and to what extent coastal flooding might happen.

The scientific tools of probabilistic thinking and data development are well adapted to strategies of dynamic design thinking for coastal resiliency. The challenges of designing for an unknown future—one that will be significantly affected by climate change—may be seen as an opportunity, particularly in urbanized coastal regions. New methods for producing resilient designs that embrace dynamic and possibly unknown novel outcomes are needed, producing research, thinking, and action that looks beyond past models into an optimistic future.

Dynamic Performance-Based Design

Innovative maps and data not only can help designers and policymakers navigate a changing climate but also can drive the development of new conceptual and institutional structures of coastal resilience. The projects described in this book conceive of the built environment as a landscape that can adjust to changing conditions rather than resist the effects of natural disasters and climate change. To make these visions possible and practical, designers and planners can develop project-specific performance standards that encourage more dynamic relationships with water. When existing codes might stymie innovation, performance-based guidelines inspire the reexamination of conventional approaches to structure, stability, and permanence. The engineering framework of performance-based design (PBD) offers a pertinent model that can be adapted to coastal resilience work. When architects and engineers design buildings to withstand wind and seismic forces, they often engage principles of PBD, transforming probability assessments—the chance that hazards of different severity will occur—into generative tools. Accordingly, the probabilistic research and mapping techniques described in this chapter also encourage PBD.

Planners and designers can use performance-based strategies to implement mitigation tactics that meet designated probability benchmarks. For example, during high-probability, low-impact events—frequent but mild storms—energy damping interventions such as wetlands, reefs, and breakwaters can reduce wave energy and moderate flooding. But during low-probability, high-impact events—rare but severe storms—wetlands, reefs, and breakwaters may be quickly subsumed by high surge. In these scenarios, seawalls or berms may prevent flooding. Alternatively, communities may need to be prepared to accept a certain amount of water by elevating houses and providing secure evacuation routes.

PBD allows architects and engineers to adapt their work to an array of probable events with respect to the function of the structure and the level of risk a property owner is willing to assume. This approach is an alternative to the more traditional code-based design, which aims to minimize damage in the event of a few narrowly defined disaster scenarios.[39] A FEMA study on seismic building standards defines PBD as a process that "explicitly evaluates how a building is likely to perform, given the potential hazard it is likely to experience, considering uncertainties inherent in the quantification of potential hazard and uncertainties in assessment of the actual building response."[40] Already an accepted method of design for seismic-, fire-, and wind-related disasters, PBD advances conventional code-based regulation in three ways that make it particularly suited to coastal resilience projects. It enables innovative design and engineering, especially adaptive structures; it protects economic interests; and it offers a framework to measure the capability of structures against projected events that may differ from historical precedents. Typical seismic design performance levels include collapse prevention, life safety, damage control, and maintained operations.[41] Because coastal resilience is regional in scale, it requires additional performance categories such as access to emergency services, evacuation routes, and environmental damage control.

PBD for Earthquake Engineering

The adoption of PBD in earthquake engineering provides a relevant model for the incorporation of a PBD approach to coastal resilience. Traditional building codes, intended to protect life safety in the event of a disaster, are prescriptive; they stipulate necessary materials, construction methods, or structural details, leaving little room for improvisation and innovation. PBD facilitates innovation by encouraging designers and engineers to develop both performance standards and the means to achieve those standards. In the aftermath of the Santa Barbara (1925) and Long Beach (1933) earthquakes, California adopted a set of codes requiring that buildings exhibit minimum levels of lateral strength.[42] Property owners came to believe that adherence to these codes would ensure safety and minimize damage during an earth tremor. However, severe damage to code-compliant buildings in the wake of the Northridge (1994) and Kobe (1995) quakes demonstrated the limitations of existing seismic design codes. Although regulatory agencies test the performance of individual codes against a range of earthquake conditions, they do not assess the performance of complete buildings. Therefore, structures designed to these specifications might perform better or worse than code compliance suggests.

The 1970s marked the beginning of a paradigm shift in structural and infrastructural design toward engineering methods that could accommodate movement and adapt to change. The 1971 San Fernando Earthquake demonstrated the fallibility of codes that stipulate only the minimum strength of rigid supports and joints and catalyzed the development of "ductile" buildings, designed to deform without catastrophic failure.[43] Rather than define the specific performance of a joint, PBD recognizes several acceptable degrees of motion depending on the strength of the earthquake. In the past decade, PBD guidelines have spurred innovation, allowing the development of high-rise towers in earthquake zones such as San Francisco and Tokyo. In 2008, a PBD system was approved as an alternative to the San Francisco's building code for structures that exceed 240 feet in height. Since then, nearly a dozen new towers have been developed using the PBD process within 30 miles of the epicenter of the 2014 South Napa Earthquake.[44] Because it allows designers to devise creative and efficient uses of methods and materials, PBD is also cost-efficient. Moreover, specific performance criteria help property owners better understand the risks that their buildings face and how these buildings might be insured or retrofitted to provide the financial or structural security that would prevent economic loss.

Most importantly, PBD offers a means to measure the capability of structures and landscapes against future events rather than past precedents. The probabilistic structure of PBD capitalizes on digital modeling to assess future hazards and their effect on buildings and sites. Before computer simulation, the most reliable metric of whether a material or structural detail could resist an earthquake was whether it survived the last seismic event. Current modeling software allows engineers to test structures against an almost infinite number of distinct, probable events. Because there are no true precedents for the intensified storm surges and increased sea levels threatening coastlines today, the use of predictive technology in the PBD process is critical.

PBD for Coastal Resilience

The implementation of PBD into regulatory structures for increasing coastal resiliency would require a three-step process. First, designers and property owners must define and describe performance objectives: how a structure, site, or system should perform in a range of flood events. Second, regulating agencies must quantify flood hazards and represent them spatially so that designers can match performance objectives to site-specific hazards. Third, engineers and scientists must develop testing methods to evaluate the degree to which a structure meets performance objectives in response to predicted hazard scenarios. Because the risk of sea level rise and storm surge will increase with climate change, PBD standards should account for change over time and be continuously updated. In this sense, coastal resilience demands *dynamic* performance-based design.

The aggregation and integration of structural and nature-based features lend themselves to evaluation by tiered PBD objectives. For example, a coastal landscape might offer multiple layers of protection from storm surge inundation—a wave-attenuating offshore breakwater followed by an artificial beach or dune and a structural barrier, seawall, or levee—that behave differently depending on the timing and severity of a storm. The performance objectives for this condition might include the following: In the case of a storm with high probability of exceedance in 50 years, there would be no damage to offshore reefs and breakwaters, limited dune and beach erosion, and no flooding behind structural barriers. For a storm with a middle-range probability of exceedance in 50 years, there would be some damage to offshore reefs and breakwaters, considerable dune and beach erosion, and a foot or less of on-land stillwater flooding behind structural barriers. And for a storm with a very low probability of exceedance in 50 years, there would be considerable damage to offshore reefs and breakwaters, complete dune and beach erosion, and several feet of flooding behind structural barriers. Such a landscape exhibits resilience across the gradient of conditions associated with moderate to extreme storms through the level of protection provided by offshore wave attenuation features and structural barriers. (See Figure 4.22.)

Many of the design projects described throughout this book exhibit strategies and systems that could be codified into a PBD framework. For example, "Water Proving Ground," LTL Architects' project for *Rising Currents*, accommodates tidal cycles and rising sea levels with buildings and landscapes that foreground water as a performative element rather than a picturesque feature. The continual motion of tidal water precipitates physical shifts in the landscape at Liberty State Park in Jersey City, New Jersey. At the proposed Aquaculture Research and Development Center, a series of floating docks rise and fall with the tide. In a low-lying park near the amphitheater, islands, pools, and beaches emerge and disappear with tidal fluctuation, eventually vanishing altogether with sea level rise. Subtle changes in elevation produce a dynamic range of contingent conditions; certain functions and landscape elements are activated only during part of the tidal cycle or specific flood

DYNAMIC PERFORMANCE
BASED DESIGN

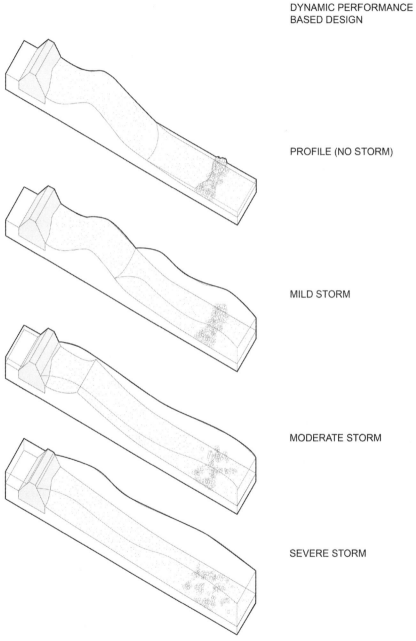

PROFILE (NO STORM)

MILD STORM

MODERATE STORM

SEVERE STORM

Figure 4.22: A dynamic
performance-based design
approach to coastal resiliency
acknowledges and quantifies
varying acceptable degrees of
storm damage based on the
severity of the storm event.

Julia Chapman, *Structures of
Coastal Resilience*, 2015

conditions. Here, *performance* holds a double meaning: The project accommodates a range of flood conditions both programmatically and structurally. Certain zones within the site can accept flooding without damaging the design, whereas other areas might need remediation after a serious flood event. Each zone within the project is calibrated to perform in distinct ways at various levels of inundation.

At a smaller scale, Harvard University's SCR project for the Hummocks in Rhode Island proposes a two-tiered system, with two sets of performance criteria. A series of layered dunes planted with shrubs will adapt and thrive despite persistent saltwater inundation on a regular basis with each high tide. But in a severe storm event, a shelterbelt of trees at a higher elevation will protect the evacuation route from Aquidneck Island. In this scenario, the shrub-covered dunes will be fully inundated for an extended period of time, and depending on the severity of the surge, potentially sacrificed.

Likewise, the atoll terraces proposed in the City College of New York's SCR project for Jamaica Bay perform an ecological function in maintaining the health of the bay while also attenuating waves and reducing wind fetch across open water during moderate storm events. These low terraces wrap the edges of the bay's marsh islands, performing ecologically by capturing sediment and helping the existing marshland migrate upward in the face of rising sea levels. Yet in a severe storm, the atoll terraces are not expected to protect the marsh islands from inundation. During severe events, other protective measures such as the verge enhancement at the Belt Parkway would be expected to protect the neighborhoods along the back-bay perimeter from surge flooding. Different landscape features thus meet varied performance criteria while working together as a layered system of protection for an expanded understanding of the shoreline.

The University of Pennsylvania's SCR proposal for Lambert's Point in Norfolk provides another example of how PBD criteria might be dynamic, shifting over time as sea level rise changes the nature of certain coastal regions. At Lambert's Point, ground will be raised incrementally over time, creating a wide spine that extends from higher ground inland out into the water. The upper ridge of Lambert's Point's high ground can initially serve immediate needs for evacuation, but with time it can develop into a more robust and dense neighborhood as citizens migrate away from particularly vulnerable low-lying areas. Nearer to the coast, low-lying areas that currently hold rainwater overflow from nearby creeks might evolve into an extension of the estuary, providing an ecological home for species that can no longer thrive in areas with increased salinity. In the context of Norfolk, where sea level rise is currently felt on a near-daily basis, this proposal for dynamic evolution is prescient and relevant. Adjusting expectations for coastal regions, as well as the performance criteria to which the built environment is held, is no longer a choice but rather a mandate.

At coastal regions, dynamic PBD also has the potential to aid the communities most vulnerable to flood damage and destruction, as well as those least able to participate in rebuilding efforts. Rising inequality threatens many coastal communities as much as rising sea levels; the most vulnerable citizens often live in the most vulnerable landscapes. Although the urban waterfront has been recently revalued and redeveloped with luxury housing for the wealthy, these low-lying coastal floodplains have historically been the site of inexpensive land for housing the poor. The ethical component of coastal resiliency and design must be considered, and the opportunity to replan coastal communities and landscapes is an imperative. In wealthy beachfront areas, the tendency to protect an individual coastal property with protective dunes or groins often runs the risk of compromising the safety of the adjacent properties by causing more erosion. Extending this phenomenon to a denser urban condition, this book supports a civic approach that addresses the design of coastal resiliency at the neighborhood or regional scale rather than at the scale of the individual home and encourages the development of ecological, infrastructural, and social resiliency. Necessary adaptation to new climate futures and the consideration of new dynamic standards of performance are imperative calls to action and present a unique opportunity to rethink, replan, and rebuild a more equitable urban coastal landscape.

Chapter 5

Centennial Projections

The impact of climate change will be transformative, quite literally changing the face of the earth as we know it. Consequentially, it provokes several recognizable responses. One response is to deny the existence of climate change and thus deny any fiscal, regulatory, or managerial responsibility. A second response is to acknowledge the expansive scale of the problem but view the necessary political challenges and economic adjustments as insurmountable. A third response is to examine isolated components of climate change through research and investigation, developing actionable projects. Most scientific research and design work operates this way, investigating a very specific aspect of climate change in a scientific laboratory or designing a single building to withstand climate forces. This way of working is necessary and important, but it also presents new challenges. In the scientific community, climate research is often published in disciplinary journals but rarely conveyed to public audiences in an in-depth way. In the design community, isolated investigations and prototypes do not often yield integrated, widespread, and cohesive interventions. Restrictions on budget, property ownership, and jurisdictional authority tend to limit the scale and scope of projects.

This book suggests an alternative response to climate change, one that encourages resilience through cross-disciplinary research and action, as well as engagement with public audiences and communities. This approach presents a shift in perspective, suggesting that the increased risk of sea level rise and storm surge hazards—both of which have the potential to drastically alter the coastline—might catalyze positive change, especially in cities. This book promotes design thinking for its imaginative capacity to reconsider the status quo and visualize an alternative future. Although the consequences of climate change may be overwhelming, the challenges ahead provide an unprecedented opportunity to build better in urban coastal cities.

The preceding chapters present an approach to coastal resilience that weaves scientific, artistic, and design practices into both a collaborative working method and a novel approach to the transformation of coastal communities. The representational techniques, design strategies, and mapping processes described here are intended to launch further research and design work. Creative design work is often the product of a collaborative and innovative process. This book is intended as a model of process as much as product. Of course, coastal resilience projects do not culminate with a finished product but rather present an evolving process shaped by people, places, and climate. The structure of this book, from site analysis to design proposals to verification through mapping, follows the developmental sequence of many of the coastal resilience projects presented in the book. This structure is intended as both reflection on the work and a guide for others.

Chapter 2 presents tools and methods for visualizing and representing the coast not as a rigid line but rather as a flexible and malleable site of exchange. Championing the ecological *informe* derived from art and environmental theory, we present techniques for interpreting the coast that capture the dynamic human and natural processes present in coastal environments. Through presentation of the representational techniques of art, architecture, and engineering, from plan and section to collage and physical models, the chapter argues for the continued expansion and interpretation of these tools to develop new methods for analyzing coastal sites and new ideas for coastal resilience.

The representational strategies described in Chapter 2 inform the design work presented in Chapter 3: coastal adaptation projects that allow the coast to be reconfigured through the harnessing of natural processes and strategic interventions. These projects, visualized at scales ranging from a reef to a neighborhood to a region and even a river delta, prioritize the principle of controlled flooding over flood control. The traditional hard engineered solutions to flood control are at risk of catastrophic failure, and new alternative visions for attenuation, planning, and protection at the coast are needed. The challenges of climate change must be addressed by not only advanced research but also compelling narratives and inspiring visions.

The techniques of adaptive resilient design described in Chapter 3 require new modes of analysis and verification in order to convince communities and authorities of their effectiveness. In response, Chapter 4 emphasizes dynamic flood mapping as a critical tool for design thinking and project evaluation, highlighting the nuances of probabilistic research and visualization that can help both communities and designers consider the possible outcomes and consequences of design interventions. Emphasizing a planning-for-hazards approach to resilience over a risk-and-replacement model, the chapter argues for maps that function as design tools rather than navigational charts, registering the dynamic nature of sites and hazards. It presents the *Structures of Coastal Resilience* mapmaking process as a model for data analysis and geospatial mapping. Flood map matrices illustrate the changing nature of storm surge hazards over the next century, as changes in storm climatology and sea level rise make rising flood depths more probable with each passing decade. The matrices also provide a model for a new conceptual structure of coastal resilience: dynamic performance-based design. This structure would allow communities to determine the levels of risk they are willing to accept and then build interventions designed to perform to flexible benchmarks.

Although this book is intended as a guide for future coastal resilience projects, it is also a call to action. Much of the work described here is speculative; such projects are important for challenging perspectives and providing alternatives to established methodological and infrastructural paradigms. Yet the transition from speculative work to pilot projects is currently under way. Projects that implement this paradigm shift in design thinking are being championed by federal agencies such as the Department of Housing and Urban Development and the U.S. Army Corps of Engineers (USACE), as well as select state and local governments.

This transition from speculative work to physical interventions must address the difficult ethical questions of social equity and environmental degradation that arise—or are revealed—with climate change. In areas that are becoming increasingly urbanized, public health and ecological health are both vulnerable to the risks associated with heat, rain, sea level rise, and surge that will accompany a warming planet. To tackle the challenges of climate change, socially progressive and environmental defense agendas must be considered in tandem; too often the most vulnerable communities are harmed the most by the effects of environmentally irresponsible development. Here, design thinking provides a framework for rethinking the relationships between people and ecology in the urban context.

Building for a Century

Much of the work described in this book, like many other resiliency propositions, is projected forward to the end of this century; for practical purposes, 2100 is a benchmark date for climate predictions as well as the final phasing of design projects. This century-long trajectory results from the research tools at hand, but it also reflects a realistic lifespan for buildings and infrastructure. Much of today's infrastructure, from subway tunnels to bridges to dams, was built a century ago. In order for the infrastructural plans developed and built today to last 100 years, this work must be done well.

The 100-year time frame is also a useful lens for looking backward. With the advantage of a century's hindsight, early twentieth-century coastal projects provide examples of both good intentions and unintended negative consequences. Coastal sites are complex, and several early twentieth-century large-scale coastal design strategies provide examples of brilliant failure. It is important to consider these projects for the lessons—both positive and negative—that they offer to contemporary planners and designers. The three projects presented here have shaped not only coastal land use but also public health, social equity, and ecological viability from the time of their implementation through the present day. Each addressed problems at the infrastructural scale of the landscape, offering large-scale interventions intended to combat considerable risks. The raising of Galveston and the construction of a massive seawall after the Great Galveston Hurricane of 1900 is one of the most significant land construction projects in American history, demonstrating an urban approach to coastal resilience but also setting the precedent of concrete seawalls as the primary means of defensive flood control on the Gulf and East Coasts. The technique of coastal salt marsh trenching in New York and New Jersey during the first decades of the twentieth century responded not to flood risk but to the contemporary understanding of disease transmission, commendably considering public health as an environmental problem but also altering the landscape profoundly in ways that ultimately threatened ecological health. Finally, the construction of public housing in New York City over the course of the twentieth century provided adequate and affordable housing for many New Yorkers, but in many cases it displaced poor communities into isolated and often low-lying regions particularly vulnerable to flooding. These public housing projects are just one example of a widespread practice of the public subsidization of homes located in floodplains, with the end result of trapping citizens in vulnerable locations that they cannot afford to leave.

These three undertakings, elaborated below, provide examples of coastal projects with isolated objectives that may exhibit good intentions yet lead to unforeseen or inadvertent consequences. Historical distance allows a perspective on useful lessons for coastal resiliency from both the positive aspirations and negative consequences of large-scale projects. These examples further illustrate that the challenges faced today are not only changes to the natural environment from climate change but also the legacy of historical decisions concerning the built environment—often the result of debates considering not only flood risk but also public health, poverty, opportunity, and social justice.

Raising Galveston

The raising of Galveston Island behind a massive new seawall was a dramatic response to an equally dramatic storm event. The elevation of the entire city was a synthetic urban strategy; instead of isolating neighborhoods from the water, entire blocks were raised to meet a new coastal condition. And yet the project's subsequent impact on future coastal risk reduction on the Gulf and East Coasts was far less urban. The USACE built similar seawalls at other locations but without the massive urban project of land raising behind them.

The hurricane that struck Galveston and the Texas Gulf Coast on September 8, 1900 was devastating. Named the Great Galveston Hurricane, the Category 4 storm caused tremendous damage to the city and killed as many as eight thousand people, making it the deadliest hurricane to hit the United States. Storm surge, estimated at more than 15 feet, was responsible for most deaths and damage. After the storm, city leaders and the USACE proposed a radical solution: raising the entire city and hardening the beachfront of the barrier island with a continuous seawall.[1] The island would essentially slope from the eastern seawall, rising to a height of 17 feet along the Gulf Coast, to the back bay on the western side of the island. The incremental process of raising the city of Galveston, which began in 1903 and took 7 years, included not only homes and buildings but also streets and services. Homes were lifted onto blocks with hydraulic jacks, and soil was then filled in underneath. Barges traveled along new canals dug specifically to bring soil and fill from Galveston Bay to elevate the city blocks. Photographs archived by the Rosenberg Library in Galveston illustrate the multiple stages of the physical construction of the elevated barrier island. (See Figures 5.1, 5.2, and 5.3.)

This approach of elevating coastlines at the ocean front, pioneered in Galveston, subsequently became a model for hurricane protection at other coastal locations along the Gulf and East Coasts of the United States. Throughout the twentieth century, the USACE has championed the use of seawalls and beach nourishment techniques to protect the front face of barrier islands and other coastal edges. Although in many storm scenarios this strategy can be effective, the emphasis on oceanfront protection often leaves barrier islands vulnerable to flooding on the back side through the entrance of surge waters at bay inlets and also subjects low-lying areas at the embayments behind barrier islands at risk. As landforms, barrier islands have evolved as a changeable geomorphology, one that provides a front line of defense to the mainland. Both seawalls and constructed elevated urban land, as in Galves-

Figure 5.1: The Great Storm of 1900 was a Category 4 hurricane that devastated Galveston Island, Texas, causing extensive property damage. Photograph by Henry H. Morris, 1900.

Courtesy of the Galveston Photographic Subject Files, Rosenberg Library, Galveston, Texas

Figure 5.2: After the 1900 hurricane, a 17-foot-high concrete seawall was constructed along the Gulf Coast at Galveston Island. Photo by Joseph M. Maurer, ca. 1904.

Courtesy of the Galveston Photographic Subject Files, Rosenberg Library, Galveston, Texas

Figure 5.3: After the completion of the seawall, the grade of the entire city of Galveston was elevated. Houses were raised on jacks or stilts and a slurry fill was pumped in through a network of pipes. Here, a view at 28th Street and Avenue P shows an area with completed fill and an area with raised houses that had not yet been filled, ca. 1908.

Courtesy of Galveston Photographic Subject Files, Rosenberg Library, Galveston, Texas

ton, anchor barrier islands in place, ultimately reducing their capacity as flexible layer of defense for the mainland behind. Moreover, this kind of construction, with its illusion of reduced risk, keeps large urban populations in highly vulnerable locations. In 2008, Hurricane Ike struck Galveston and the Texas coast. The higher front face of Galveston Island at the seawall suffered limited flooding, but surging floodwaters reached a height of 10 feet in some locations on the back side of the island facing Galveston's West Bay.

Looking forward to the next one hundred years with a vision for a resilient coast, the lessons learned from the seawall approach to flood infrastructure at the barrier island must be reconsidered. The historical emphasis on the oceanfront as the line of protection against mainland flooding has long-term flaws. Stabilization of the shoreline through the infrastructural mechanisms of seawalls and beach nourishment has proven to be both unsustainable and fallible. Beach nourishment—the technique of elevating beaches with massive volumes of sand—is an expensive procedure that must be repeated every few years, as littoral currents and storms repeatedly erode and wash away this valuable sand fill. The engineered seawall produces a higher elevation at the coast but has resulted in higher risk for communities on the low-lying embayments behind the barrier islands. The coastline must be accepted as a dynamic and changeable system, not as a line to be fixed in perpetuity; this engineered attempt to enforce stasis at the coast will ultimately fail. As flood infrastructure is reenvisioned and developed for the next one hundred years, a new approach is necessary, one that adapts to the scale of the landscape and allows this landscape to perform dynamically as an attenuating feature for storm risk reduction.

Marsh Trenching

Whereas the raising of Galveston's elevation and the construction of its seawall were a direct response to a storm event, other historic large-scale landscape transformations were implemented because of public health concerns. Widespread marsh trenching in New York and New Jersey in the early twentieth century illustrates a condition of public health concerns transforming the landscape but inadvertently increasing storm risk, both despite and because of shifts in an understanding of disease transmission. Marsh trenching or ditching resulted from an admirable consideration of public health and environmental health as associated concerns, yet an inadequate understanding of wetland ecology led to ecological harm.

Wetland drainage in response to the spread of disease has roots in the ancient Greek theory of miasma, that disease was spread through noxious air that emerged as a poisonous vapor from damp soil. Moist earth—swamps, marshes, wetlands—was viewed as the source of illness and therefore needed draining and reappropriation. By the mid-nineteenth century, the miasma theory—a precursor to the contagionist or germ theory—took hold in Europe and the United States. Many miasmists emphasized the role of atmosphere and environment in the spread of disease between people, viewing dry land as healthy ground in contrast to the "unhealthy" terrain of wetlands. Gentleman farmers versed in agricultural engineering, such as George E. Waring and Frederick Law Olmsted, applied the skills developed for draining agricultural fields via subsurface ceramic tile drains to urban parks. In densely populated cities, the new field of public sanitary engineering emerged. Municipal authorities sought to avoid the spread of cholera, typhoid, malaria, and yellow fever. Consequentially, wetlands and lowlands were drained in the name of public health in Europe and the United States, and vast nineteenth-century public spaces such as Central Park in New York were transformed into sanitary mechanical scrubbers, quickly shedding and transporting surface water and stormwater. Fortuitously, the nineteenth-century miasmists were successful in stemming the spread of disease—not by preventing the earth's noxious exhalations of "bad air" that they believed to be the cause of disease transmission but by eliminating the presence of stagnant water. Though unknown to the miasmists, the real disease culprit was not the damp earth but the stillwater habitat of a germ-infected vector: the *Aedes* mosquito.

After huge losses of U.S. Army troops to yellow fever in Havana during and after the Spanish–American War of 1898, army physician Walter Reed successfully proved in August 1900 that the disease was spread by a bite from the *Aedes aegypti* mosquito carrying the yellow fever virus.[2] This mosquito, now commonly known as the yellow fever mosquito, is also the main carrier of today's Zika virus. Other current mosquito-borne diseases include malaria, the West Nile virus, dengue fever, chikungunya, and several types of encephalitis.

Paradoxically, the advent of germ theory and the identification of the mosquito as a disease vector in 1900 led to massive swamp draining projects along the East Coast of the United States, resulting in marsh fragmentation and demise.[3] Extensive wetland "mosquito trenching," or the cutting of V-shaped linear open ditches, rapidly drained—and fragmented—

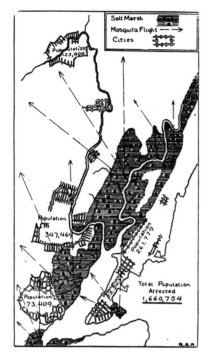

Figure 5.4: "How the salt-marsh mosquito operates," a map from Thomas J. Headlee's *The Mosquitoes of New Jersey and Their Control*, 1921, indicating the threat of salt marsh mosquitoes to the populations of Paterson, Newark, Jersey City, and Elizabeth. The map was used to support extensive marsh drainage projects in New Jersey.

Rutgers/The State University of New Jersey

intertidal marshes. Mosquito ditches were first implemented at New York City's marshes at Jamaica Bay, Coney Island, and along the southern shore of Staten Island in the early twentieth century. Like George E. Waring's underground tile drainage of Central Park, this marsh trenching created dry ground by whisking away surface water, with the unintended consequence of causing serious fragmentation and degradation of massive salt marsh complexes, destroying acres of coastal wetland salt marshes and compromising the integrity and many ecological benefits of the remaining marsh complexes. (See Figures 5.4, 5.5, 5.6, and 5.7.) By 1919, the state of New Jersey reported trenching approximately 120,000 acres of salt marsh, which involved the cutting of 18,244,217 linear feet of open ditches (10 inches wide and 24 to 30 inches deep) to destroy mosquito breeding habitats.[4]

Although driven by a real and substantial fear of mosquito-borne disease, these drainage projects were based on a fundamental misunderstanding of marshland and mosquitoes. Indeed, ponded standing water does provide habitat for mosquito larvae, but intertidal action moves water through healthy and robust wetlands that in turn provide the valuable habitat that supports other species that feed on mosquito populations. The best management of mosquitoes is not draining saltwater marshes but rather maintaining a robust intertidal system. Low tides drain mosquito breeding sites. Dynamic water movement between high and low tides also reduces harmful bacteria and improves water quality for plant and animal populations. The ground need not be bone dry to prevent mosquito breeding; however, it must support the movement and infiltration of water.

Protecting environments and people from mosquito-borne illness remains a complex task. The recent outbreak of the Zika virus in both North and South America represents another type of risk from climate change and the consequences of migrating disease vectors. Today, international health officials struggling to contain the spread of the Zika virus advise citizens to remove or cover containers or rain gutters that might collect standing water. Spraying larvicide over nonresidential areas, which has environmental complications, is not a particularly effective tactic against *Aedes aegypti* because this species breeds primarily in gardens and homes, but it is a widely practiced technique. Health risks and unintended environmental consequences must be carefully considered by health officials, landscape architects, and environmental advocates.

The marsh trenching and draining of the last century have had a lasting and pervasive impact on the coastal landscape. Many former wetlands

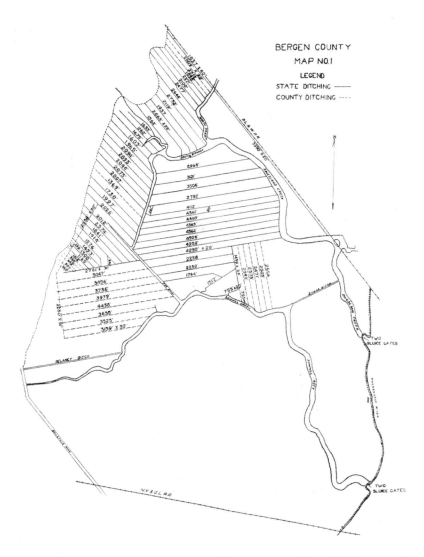

Figure 5.5: Map of a ditching system for Bergen County, New Jersey. Each proposed ditch was 10 inches wide and 30 inches deep. Drawing from Thomas J. Headlee's *The Mosquitoes of New Jersey and Their Control*, 1921.

Rutgers/The State University of New Jersey

have been filled and developed with built structures, although they remain low-lying territories prone to coastal flooding. A lack of knowledge of the ecological benefits of marshes devalued these wetland territories. Yet the benefits of marsh complexes are now understood to be manifold. A healthy marsh ecosystem provides many ecosystem services, including water purification, groundwater recharge, and carbon sequestration. Wetlands provide valuable habitat for fish and wildlife, including several endangered species. In addition, marshes have the capacity to absorb water during flood events and

Figure 5.6: Aerial view of Jamaica Bay's JoCo Marsh near John F. Kennedy International Airport, with the straight lines of early twentieth-century mosquito ditching visible at the bottom left of the image.

Photo © 1989 Don Riepe/American Littoral Society

attenuate waves during storms, reducing shoreline erosion and the severity of flood impact. As coastal resiliency design work is developed, the impact of this work should be projected one hundred years forward, ideally producing the conditions for future wetlands to thrive. Robust wetland systems would significantly benefit and support a new resilient coastal landscape.

Public Housing in the Floodplain

Although they may not have fully understood disease transmission, both the miasmists and contagionists recognized the relationship between public health and the environment. The same may be said for early- and mid-twentieth-century advocates of public housing in New York City. An alternative to overcrowded tenement neighborhoods, public housing was championed by its supporters as providing an opportunity for clean, safe, and affordable living spaces for workers in an expensive city where housing was often in short supply. Some public housing was built on centrally located cleared land, but much of it was built on vacant or previously industrial land, often at the very outskirts of New York City. In many cases, this land was—and remains—low-lying and vulnerable to flooding.

Three public housing projects in Brooklyn and Queens provide case study examples of the circumstances through which flood-prone land became home to low-income New Yorkers, to deleterious effect. Completed in 1939, the Red Hook Houses in southwestern Brooklyn were funded by a new federal public housing system created in 1937, led by Nathan Straus Jr. They were built by the New York City Housing Authority (NYCHA), led at the time by Alfred Rheinstein, an appointee of mayor Fiorello La Guardia. Rheinstein was an efficient and economic builder, and he favored building new housing on marginalized industrial land to escape the challenges and expenses of building in Lower Manhattan.[5] The Red Hook Houses remain one of the largest public housing developments in New York City, home to six thousand people. The neighborhood is isolated from public transportation and physically disconnected from the rest of Brooklyn by the elevated Brooklyn–Queens Expressway. The low-lying, formerly industrial waterfront land proved disastrous for its residents during the 2012 landfall of Hurricane Sandy. Nearly the entire extent of the Red Hook neighborhood was flooded, with the Red Hook Houses experiencing a lack of power and running water for weeks after the storm.

NYCHA's Gowanus Houses were built on former swampland adjacent to the Gowanus Canal in the Gowanus neighborhood of Brooklyn. As early as the 1830s, land speculators drained the marshland and infilled the lowlands to sell off land at a profit. Though filled, the land remained at a low elevation and highly vulnerable to flooding from both the waterfront and the canal. Completed in 1949, the Gowanus Houses consist of fourteen buildings on 13 acres of land, containing more than one thousand units that house approximately three thousand people. Drawings from the design of the Gowanus Houses in the 1940s affirm that the project team was well aware of the vulnerability of the site. A contour plan calls out the demarcations of an "old peat bog," and a small plan on a sheet of cross-sections of the site designates a large swath of "swamp." (See Figures 5.8, 5.9, and 5.10.) During Hurricane Sandy, floodwaters damaged below-grade electrical rooms and destroyed the complex's boiler. Residents remained without power for almost 2 weeks, although surrounding homes and businesses never suffered any loss of utilities. Elderly residents on upper floors were left stranded without functioning elevators.

During the postwar era, New York City acquired, through purchase or eminent domain, large low-cost building parcels along coastal lands at the Rockaway Peninsula and Coney Island—at the far reaches of the boroughs of Brooklyn and Queens—with the intention of "improving" these parcels with public housing developments. Robert Moses, who created and held sway over several public authorities including NYCHA in the 1940s and 1950s, favored building on public lands, often in undesirable areas far from the

Figure 5.7: View of the "Manahan power ditcher" used to cut drainage ditches in New Jersey marshland for the intended purpose of eliminating mosquito breeding grounds. Photograph from Thomas J. Headlee's *The Mosquitoes of New Jersey and Their Control*, 1921.

Rutgers/The State University of New Jersey

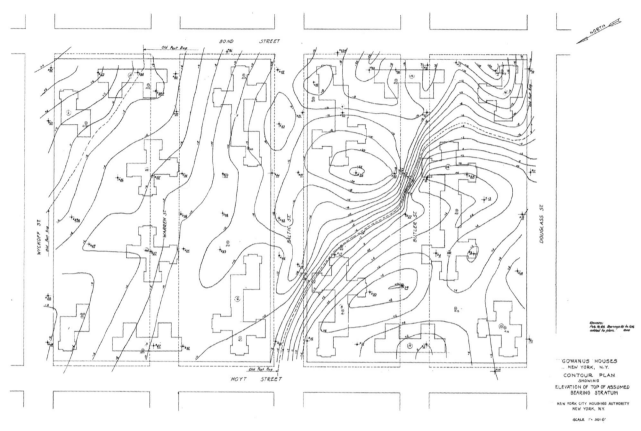

Figure 5.8: Contour plan of bearing stratum, Gowanus Houses, Brooklyn. A New York City Housing Authority project initiated in 1948, the plan denotes the extent of an "old peat bog," a low-lying wetland prone to flooding.

New York City Housing Authority

centers of real estate development in Manhattan. Both the Rockaways and Coney Island were low-density summer beach communities in decline in the postwar period, and Moses sought to transform these areas into year-round higher-density residential centers, arguing for urban renewal in the name of health, sanitation, and housing improvement.[6] With the construction of multiple housing projects at these coastal areas, the former summer bungalow neighborhoods became home to large populations of low-income NYCHA tenants displaced from the denser areas of the city and often poorly served by public transportation. Perhaps unwittingly, the strategy established a significant density of isolated low-income housing in areas particularly vulnerable to the risk of the ocean's storm surges and flooding.

Low-lying public housing in New York City is consistent with a long history in the United States of housing poor communities in the less valuable real estate of low-elevation floodplains. When prompted or forced by federal regulation to build low-income housing, municipalities have often acquired inexpensive marginalized lands to build housing stock for poor communities, providing "improved" housing but in tenuous or marginalized locations, often flood-prone. An earlier example of inequity in housing, the 1927 Mississippi River flood was particularly devastating to poor African American sharecroppers who farmed land adjacent to and below the levees. But the practice of encouraging vulnerable communities to remain in vulnerable locations continues today. The National Flood Insurance Program (NFIP) allows federal subsidization of homeownership in floodplains, encouraging permanent settlement and resettlement in areas that flood repetitively. Because their subsidized flood insurance rates to the NFIP remain low and disproportionate to actual flood risk, homeowners often have no choice but to rebuild in place, unable to afford to move to higher ground.

As new design thinking develops and transforms the coastal landscape, a resilient vision must consider not only the terrain and ecologies but also the landscapes of social and environmental justice for those who inhabit these territories. Resilient landscapes must also be equitable landscapes. New thinking regarding housing in the urban realm should address past inequities, support social resilience, and provide opportunities and access for all citizens.

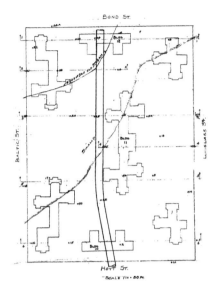

Figure 5.9: A key plan of the Gowanus Houses on a drawing of cross-sections taken through surveyed vertical borings denotes the large swath of "swamp" in the middle of the public housing site.

New York City Housing Authority

Looking Forward

The large-scale infrastructural projects of the twentieth century presented here demonstrate that a lasting impact—whether positive or negative—is possible with sufficient political will. The current era demands a new kind of courage and strength, not only for building new infrastructure but also for managing retreat from particularly vulnerable areas. Once seen as politically infeasible, managed retreat is being increasingly studied in locations where sea level rise threatens to make "nuisance" flooding even more frequent and life-threatening. Communities and policymakers are beginning to consider the possibility of retreat from the coast, how it might be implemented in a fair and equitable way, and what it would look like. Coastal retreat requires

Figure 5.10: View of the Gowanus Houses under construction at the Brooklyn site, bounded by Hoyt, Bond, Douglass, and Wyckoff Streets, 1949.

Brooklyn Collection, Brooklyn Public Library

not only upland densification but also careful consideration of the management of the landscapes left behind. Planning for retreat requires reflection on the interconnected questions addressed by advocates of social and environmental justice, so that the retreat is not perceived as a strategy of involuntary displacement. As this book demonstrates, robust coastal resilience must interweave both resilient communities and resilient ecologies.

In the face of climate change and its tremendous challenges, a new approach to coastal resilience offers an opportunity for a radical and equitable transformation of the landscape. Through collaborative design thinking, a new vision for coastal cities is possible, one that is adaptive and resilient. Looking forward one hundred years, it is possible to envision and design an urban future that is not only resilient given the impacts of climate change, but also produces a more equitable social and ecological landscape capable of thriving in this unknown future climate.

Afterword

Jeffrey P. Hebert

VICE PRESIDENT FOR ADAPTATION AND RESILIENCE,
THE WATER INSTITUTE OF THE GULF

To say the 2017 Atlantic hurricane season has been active would be a terrible understatement. In September, Hurricane Harvey—the first major Category 3 or higher hurricane to hit the United States since the disastrous 2005 Atlantic season—came ashore along the southeastern Texas coast. What quickly became a Category 4 storm dropped precipitation of biblical proportions across a broad swath of Texas and Louisiana. The ensuing 50+ inch deluge flooded Houston, the fourth largest city in the United States and a primary economic hub of the country. As if Harvey weren't enough, just days later Hurricane Irma gained Category 5 classification and soon after caused widespread devastation across the Caribbean and the Florida Keys. A few weeks after Hurricane Maria's devastation in late September, Puerto Rico and the Virgin Islands remained in response mode while a fast-moving Hurricane Nate swept through the northern Gulf Coast with a near miss to my home of New Orleans. That is a great deal of devastating tropical cyclone activity across the coastal United States for just one summer.

What is so profound about the 2017 hurricane season are the weaknesses that these tropical events continue to expose. Similar to Hurricane Katrina's ability to reveal the frayed physical and social infrastructure of New Orleans to the world, Hurricane Harvey laid bare our disregard for sustaining natural systems in an aim to accommodate growth and development. This is not new, but unfortunately it is a continuing pattern in the United States. The impact of these storms should serve as a wakeup call for rapidly urbanizing delta and coastal cities across the world. The sheer volume of tropical activity during the summer of 2017 should be a warning to everyone that the risks are real and that communities need support for proactive, integrated adaptation infrastructure and innovative approaches that reduce risk in an era of increasing vulnerability. We should no longer have to react to the emergency situations and doomsday scenarios that are already reality. During Hurricane Irma, bays and rivers in south Florida and the Tampa Bay region were subjected to reverse storm surge, the rare phenomenon where those waters were literally sucked into the sea. Predictions of increased storm surge resulting from that phenomenon sent panic throughout communities across Florida.

Across the world, coastal cities are grappling with the effects of climate change and sea level rise that threaten our people, our environment, and our way of life. In New Orleans, we have seen an increase in precipitation (the summer of 2017 was the wettest on record), an increase in the level of the sea as measured at the Lake Pontchartrain seawall, and increasing subsidence of the very land beneath our feet. All these factors have prompted the development of a multitiered strategy to save this semiaquatic, deltaic city. How we should do so is not a mystery. For New Orleans to be resilient and survive, we have to continually adapt the city to its ever-changing environment. The coast must be restored to provide the natural protection the city has enjoyed for much of its 300-year history. The flood defenses and surge barriers must be maintained and strengthened. The urban environment must adapt to manage additional precipitation, and our citizens must learn to live with water.

Today there is a wealth of data available, along with dedicated professionals producing probabilistic models as tools that communities can use to inform the urban form of the future. By acknowledging vulnerability and analyzing the geospatial impacts on people and property, communities can begin to prepare, mitigate, and adapt to these changes.

Returning to Hurricane Harvey, there is very little evidence that the deluge that inundated Houston during Harvey could have been managed by any city. However, what we do know is that unbridled development into the natural drainage infrastructure of Houston's landscape has increased its vulnerability to rain events and probably exacerbated the flooding. No doubt a "can do" city, Houston will begin to reassess not only its development patterns but also its infrastructure in order to cope with what has become the new normal. Houston has experienced numerous flood events over the past few years.

After Hurricane Katrina, New Orleans began to rethink how it lives with water. Heavily influenced by Dutch thinking, city stakeholders created the Greater New Orleans Urban Water Plan to address the interior drainage system of the city and its suburbs. This plan, which integrates the massive pumping, canal, and flood wall systems with nature-based solutions, intends to add capacity for water storage across the city—while acknowledging that the city's extensive pumping system cannot provide the projected needs for increased capacity. The Urban Water Plan also harnesses the natural drainage patterns of the city as a tool to reduce reliance on costly, energy-dependent pumping. The city's resilience strategy, "Resilient New Orleans," builds on the Urban Water Plan and promotes an "Adapt to Thrive" approach, which contemplates a "Curb to Coast" risk reduction system, linking the viability of the city to the health of its coast and the management of water in the city.

The New Orleans case is a good guide for other cities as they begin transforming the relationship between the built and natural environments. Postwar development in New Orleans eliminated much of the existing marshlands that formed the natural drainage system for the over 200-year-old city. And the "new" form of that development was in stark contrast to the history of development in the city. Elevated houses with high ceilings for ventilation, clustered in dense neighborhoods on naturally high ground, were abandoned for sprawling slab-on-grade suburban developments—replacing natural flood protection with sprawling buildings in seas of concrete. The infrastructure failure of Katrina was the wakeup call and the turning point for New Orleans.

As Seavitt, Nordenson, and Chapman suggest, the future of coastal cities must be dynamic, changing, and ever evolving. Cities must constantly adapt to thrive. From Bangkok to Surat to Melbourne, cities will depend more and more on probabilistic science and amphibious city design. To be resilient, these new approaches must also provide multiple benefits to communities. Projects that adapt must provide for accessible green space and recreation space, especially in communities where these amenities are scarce. These projects must also explicitly link to equity goals and economic opportunity for vulnerable populations that are too often ignored or disconnected from the emerging blue/green economy.

The future of coastal communities lies in being prepared to adapt to ever-changing conditions. That future is already here.

Endnotes

Chapter 1

1. Manfredo Tafuri, *Venice and the Renaissance*, trans. by Jessica Levine (Cambridge, MA: MIT Press, 1989), 154–5.
2. Todd Shallat, *Structures in the Stream: Water, Science, and the Rise of the US Army Corps of Engineers* (Austin: University of Texas Press, 1994).
3. In her 2014 book *The Resilience Dividend*, former president of The Rockefeller Foundation Judith Rodin outlines these origins.
4. Lance Gunderson and Craig Allen, "Introduction," in *Foundations of Ecological Resilience,* ed. Land Gunderson, Craig Allen, and C.S. Holling (Washington, DC: Island Press, 2009), xvi.
5. C.S. Holling, "Engineering Resilience versus Ecological Resilience," in *Foundations of Ecological Resilience*, ed. Land Gunderson, Craig Allen, and C. S. Holling (Washington, DC: Island Press, 2009).

Chapter 2

1. Claes Oldenburg quoted in Richard Axsom and David Plaztker, *Printed Stuff: Prints, Posters and Ephemera by Claes Oldenburg, A Catalogue Raisonée 1958–1996* (New York: Hudson Hills Press, 1997), 177.
2. Yve-Alain Bois and Rosalind Krauss, *Formless, A User's Guide* (New York: Zone Books, 1997), 18.
3. William Cronon, "The Trouble with Wilderness: Or, Getting Back to the Wrong Nature," in *Uncommon Ground: Rethinking the Human Place in Nature* (New York: Norton, 1996), 23–90.
4. Leo Marx, *The Machine in the Garden: Technology and the Pastoral Ideal in America* (Oxford: Oxford University Press, 1967), 353.
5. Raymond Williams, "Ideas of Nature," in *Problems in Materialism and Culture: Selected Essays* (New York, London: Verso, 1980), 67.

6. Robert Smithson, "A Tour of the Monuments of Passaic, New Jersey," in *Robert Smithson: The Collected Writings*, ed. Jack Flam (Berkeley: University of California Press, 1996), 71–72.

7. Ibid., 74.

8. Robert Smithson, *Rundown*, 1969, a film directed by Robert Fiori, with voiceover by Nancy Holt. Film created posthumously in 1993 with drawings, stills, and footage relating to Smithson's "pours" in Chicago, Vancouver, and Rome.

9. Ibid.

10. Robert Smithson, "Frederick Law Olmsted and the Dialectical Landscape," in *Robert Smithson: The Collected Writings*, ed. Jack Flam (Berkeley: University of California Press, 1996), 164.

11. Anne Whiston Spirn, "Constructing Nature: The Legacy of Frederick Law Olmsted," in *Uncommon Ground: Rethinking the Human Place in Nature*, ed. William Cronon (New York: Norton, 1996), 91.

12. Ibid., 112.

13. John McPhee, "Atchafalaya," in *The Control of Nature* (New York: Farrar, Straus and Giroux, 1980), 5.

14. Harold N. Fisk, "Geological Investigation of the Alluvial Valley of the Lower Mississippi River," conducted for the Mississippi River Commission (Vicksburg, MS: 1944), 37.

15. Bruno Munari, *The Sea as a Craftsman* (Mantova, Italy: Maurizio Corraini, 1995).

16. "Jamaica Bay Pen Project," last modified June 2014. http://jamaicabaypens.tumblr.com/.

17. Denis Cosgrove, "The Measures of America," in *Taking Measures across the American Landscape*, ed. James Corner and Alex S. MacLean (New Haven, CT: Yale University Press, 1996), 4.

18. James Corner, "Aerial Representation," in *Taking Measures across the American Landscape*, 16.

19. Lyman P. Van Slyke, *Yangtze: Nature, History, and the River* (New York: Addison-Wesley Publishing Co., 1988), 23.

20. The name *Ohmsett* is an acronym for "Oil and Hazardous Materials Simulated Environmental Test Tank." Originally built in 1974 by the U.S. Environmental Protection Agency (EPA), the tank is now owned by the U.S. Navy.

21. LiDAR is an acronym that, according to NOAA, stands for "Light Detection and Ranging." Other sources claim the term combines "light" and "radar."

22. Rosalind Krauss, "Grids," in *The Originality of the Avant-Garde and Other Modernist Myths* (Cambridge, MA: MIT Press, 1986), 9.

23. "Hilla Becher in Conversation with Thomas Weaver," *AA Files* No. 66 (2013): 17–36.

24. Edoardo Salzano, ed., *An Atlas of Venice: The Form of the City on a 1:1000 Scale Photomap and Line Drawing* (New York: Princeton Architectural Press, 1989), 9–45.

25. Linda Pollak, "Matrix Landscape: Construction of Identity in the Large Park," in *Large Parks*, ed. Julia Czerniak and George Hargreaves (New York: Princeton Architectural Press, 2007), 87–119.

26. Ian L. McHarg, *Design with Nature* (Garden City, NY: American Museum of Natural History/Natural History Press, 1969), 144.

27. *Coastal Risk Reduction and Resilience: Using the Full Array of Measures.* US Army Corps of Engineers, Directorate of Civil Works, September 2013, 3.

28. Aby Warburg, *Mnemosyne Atlas.* Cornell University Library, The Warburg Institute, Cornell University Press, 2013. http://warburg.library.cornell.edu/about

29. Charles Waldheim and Andrea Hansen, eds., *Composite Landscapes: Photomontage and Landscape Architecture* (Berlin: Hatje Cantz, 2013), 58.

30. Paul Edwards, *A Vast Machine* (Cambridge, MA: MIT Press, 2013), xiii.

31. Glenn D. Lowry, "Foreword," in *Rising Currents: Projects for New York's Waterfront*, ed. Barry Bergdoll (New York: Museum of Modern Art, 2011), 7.

32. Bartholomew Price, "Section 8.—Newton's Principle of Similitude," in *A Treatise on Infinitesimal Calculus: The Dynamics of Material Systems* (Oxford: University Press, 1889), 168.

33. Bernard Le Mehaute and Daniel M. Hanes, *The Sea: Ocean Engineering Science* (New York: J. Wiley & Sons, 1990), 959.

34. Kenneth J. Hsu, *Physics of Sedimentology: Textbook and Reference*, 2nd ed. (Berlin: Springer, 1989), 105.

35. Willi H. Hager, "Fargue, Founder of Experimental River Engineering," *Journal of Hydraulic Research* 41:3 (2003): 227–33.

36. Osborne Reynolds, "An Experimental Investigation of the Circumstances Which Determine Whether the Motion of Water in Parallel Channels Shall Be Direct or Sinuous and of the Law of Resistance in Parallel Channels," in *Proceedings of the Royal Society of London* 35:224–226 (1883): 84–99.

37. Ben H. Fatherree, "The First 75 Years: History of the Hydraulics Engineering at the Waterways Experiment Station," US Army Corps of Engineers, Engineer Research and Development Center, 2004.

38. Ibid.

39. Kristi Dykema Cheramie, "The Scale of Nature: Modeling the Mississippi River," in *Places Journal*, March 2011. https://placesjournal.org/article/the-scale-of-nature-modeling-the-mississippi-river/

40. Guy Nordenson, Catherine Seavitt, and Adam Yarinsky, *On the Water: Palisade Bay* (Ostfildern, Germany: Hatje Cantz, Verlag/MoMa, 2010), 84.

41. Cheramie, "The Scale of Nature."

42. Nordenson et al., *On the Water,* 84.

43. Ben Fatherree, "The First 75 Years."

44. Cheramie, "The Scale of Nature."

45. "Hydraulic Sediment Response Modeling," USACE: Applied River Engineering Center, St. Louis District, http://mvs-wc.mvs.usace.army.mil/arec/HSR_Model.html

46. Ibid.

47. Dmitris Stagonus, "Micro-Modelling of Wave Fields," Dissertation (University of Southampton, School of Engineering and the Environment, 2010).

48. Bahman Zohuri, *Dimensional Analysis and Self-Similarity Methods for Engineers and Scientists* (Cham, Switzerland: Springer International Publishing, 2015).

49. "Physical Hydraulic Modeling," last modified November 11, 2004, http://www.usbr.gov/pmts/hydraulics_lab/services/hydmodeling.html

50. Sultan Alam, "Let Us Try to Save the Vanishing Mississippi River Delta," in *Journal of the Louisiana Section of the American Society of Civil Engineers* 17:2 (February 2009), http://www.lasce.org/documents/journal/2009-02.pdf

51. Ashley Arceneaux, "Water Campus: LSU's Commitment to the Coast Movement Gains Traction with New River Model Facility," https://sites01.lsu.edu/wp/lovepurple/2013/12/19/lsus-committed-to-the-coast-movement-gains-traction-with-new-river-model-facility/

52. Ashley Berthelot, "Modeling the Mississippi: LSU Researchers Study Methods to Use River Sediment to Repair the Coast," http://www.lsu.edu/departments/gold/2011/05/mississippi.shtml; Alam, "Let Us Try to Save the Vanishing Mississippi River Delta"; Ernie Ballard, "River Studies: Center for River Studies to Provide Venue for Multi-Discipline Collaboration and Coastal Research."

53. Catherine Seavitt, "Yangtze River Delta Project," in *Scenario 03: Rethinking Infrastructure*, Spring 2013, http://scenariojournal.com/article/yangtze-river-delta-project/

Chapter 3

1. Gilbert F. White, "Human Adjustment to Floods: A Geographical Approach to the Flood Problem in the United States," Dissertation (The University of Chicago, Department of Geography, 1945), 2.

2. See Gregory Squires and Chester Hartman, eds., *There's No Such Thing as a Natural Disaster: Race, Class, and Hurricane Katrina* (New York: Routledge, 2006).

3. John Barry, *Rising Tide: The Great Mississippi Flood of 1927 and How It Changed America* (New York: Simon & Schuster, 1998).

4. Sherwood Gagliano was one of the first to document the rapid loss of wetlands on the Louisiana coast and propose major diversions off the Mississippi in the late 1960s and early 1970s. Mike Tidwell, *Bayou Farewell: The Rich Life and Tragic Death of Louisiana's Cajun Coast* (New York: Vintage, 2003).

5. U.S. Army Corps of Engineers (2013) Coastal Risk Reduction and Resilience. CWTS 2013-3. (Washington, DC: Directorate of Civil Works, U.S. Army Corps of Engineers).

6. Ibid.

7. U.S. Army Corps of Engineers, "North Atlantic Coast Comprehensive Study" (2015): 7.

8. See http://coastal.la.gov/ for a full summary of Louisiana's CPRA and the ongoing Coastal Master Plan. Note that the financial support of this master plan is derived from settlement fines paid by British Petroleum for the 2010 Deepwater Horizon oil spill.

9. See LOLA landscape architects, *Dutch Dikes* (Rotterdam: NAi Publishers, 2014): 30.

10. Tracy Metz and Maartje van den Heuvel, *Sweet and Salt: Water and the Dutch* (Rotterdam: NAi Publishers, 2012): 123–126.

11. Guy Nordenson, Catherine Seavitt, and Adam Yarinsky, *On the Water: Palisade Bay* (Ostfildern, Germany: Hatje Cantz Verlag/MoMA, 2010).

12. See 2009 conference and subsequent report: Douglas Hill, Malcolm Bowman, and Jagtar Khinda, eds., *Against the Deluge: Storm Surge Barriers to Protect New York City* (Reston, VA: American Society of Civil Engineers, 2013).

13. David Dunlap, "City Rolls Out a Rocky Welcome Mat for Mussels," in *New York Times*, July 27, 2012.

14. See New York district projects: http://www.nan.usace.army.mil/Missions /CivilWorks/ProjectsinNewYork.aspx

15. Robert France, *Wetland Design: Principles and Practices for Landscape Architects and Land-Use Planners* (New York: Norton, 2002), 13.

16. See also U.S. Army Corps of Engineers, "Chelsea Heights, Atlantic City, Atlantic County, NJ," Philadelphia District/Marine Design Center, May 2015.

17. U.S. Census Bureau, *2010 Census*. http://www.census.gov/2010census/data/; Trulia, LLC, *Trulia Maps*, https://www.trulia.com/local/atlantic-city.nj

18. Geohazards International, "Designing and Constructing a Tsunami Evacuation Park," viewed June 18, 2015, http://geohaz.org/projects/sumatra_tep.html

19. Jack Eggleston and Jason Pope, "Land Subsidence and Relative Sea-Level Rise in the Southern Chesapeake Bay Region" (Washington, DC: U.S. Department of the Interior, U.S. Geological Survey): 1.

20. Source of this projection is the Intergovernmental Panel on Climate Change–selected RCP 8.5. Robert E. Kopp, Radley M. Horton, Christopher M. Little, Jerry X. Mitrovica, Michael Oppenheimer, D. J. Rasmussen, Benjamin H. Strauss, and Claudia Tebaldi, "Probabilistic 21st and 22nd Century Sea-Level Projections at a Global Network of Tide-Gauge Sites," *Earth's Future* (2014): 383–406.

21. Norfolk plans, https://www.norfolk.gov/DocumentCenter/View/16292

22. Anuradha Mathur and Dilip da Cunha, *Mississippi Floods: Designing a Shifting Landscape*, (New Haven, CT: Yale University Press, 2001).

23. See Robert R. Twilley et al., "Co-evolution of Wetland Landscapes, Flooding, and Human Settlement in the Mississippi River Delta Plain," *Sustainability Science* 11, no. 4 (July 2016): 711–731.

24. Zhijun Ma, David S. Melville, Jianguo Liu, Ying Chen, Hongyan Yang, Wenwei Ren, Zhengwang Zhang, Theunis Piersma, and Bo Li, "Rethinking China's New Great Wall," *Science* 346 (2014): 912–4.

Chapter 4

1. Kerry Emanuel, *Divine Wind* (New York: Oxford University Press, 2005), 147.

2. Ian Hacking, *The Taming of Chance* (Cambridge: Cambridge University Press, 1990), 2.

3. Mark Fackler, "Tsunami Warnings, Written in Stone," April 20, 2011.

4. Lewis Carroll, *The Hunting of the Snark* (London: Macmillan, 1876).

5. National Research Council, Water Science and Technology Board, Board on Earth Sciences and Resources/Mapping Science Committee, Committee on FEMA Flood Maps, *Mapping the Zone: Improving Flood Map Accuracy* (Washington, DC: National Academies Press, 2009), 14.

6. Scott Gabriel Knowles and Howard C. Kunreuther, "Troubled Waters: The National Flood Insurance Program in Historical Perspective," in *Journal of Policy History* 26:3 (2014): 334.

7. Ibid., 333.
8. Ibid., 333.
9. Ibid., 327.
10. National Research Council et al., *Mapping the Zone*, 17.
11. Sarah Jo Peterson, "An Unflinching Look at Flood Risk," in *Urban Land: The Magazine of the Urban Land Institute*, November 21, 2014. http://urbanland. uli.org/sustainability/unflinching-look-flood-risk/
12. National Research Council et al., *Mapping the Zone*, 21.
13. Erwann O. Michel-Kerjan, "Catastrophe Economics: The National Flood Insurance Program," in *The Journal of Economic Perspectives* 21:4 (Fall 2010): 177.
14. Al Shaw, Theodoric Meyer, and Christie Thompson, "Federal Flood Maps Left New York Unprepared for Sandy—and FEMA Knew It," in *Pro Publica*, December 6, 2013. http://www.propublica.org/article/federal-flood-maps -left-new-york-unprepared-for-sandy-and-fema-knew-it
15. Sarah Jo Peterson, "An Unflinching Look at Flood Risk."
16. FEMA, "Map Modernization." https://www.fema.gov/map-modernization
17. National Research Council et al., *Mapping the Zone*, 1.
18. Ibid., 5.
19. FEMA, "Region II Coastal Analysis and Mapping," http://www.region2 coastal.com/
20. The Biggert–Waters Flood Insurance Reform Act requires FEMA to consider using the "best available science regarding future changes in sea levels, precipitation, and intensity of hurricanes" in its flood maps, http://inside climatenews.org/news/20130204/climate-change-global-warming-flood-zone- hurricane-sandy-new-york-city-fema-federal-maps-revised-sea-level-rise
21. NOAA's Sea Level Rise Planning Tool, Georgetown Climate Center, Georgetown Law, June 2013, http://www.georgetownclimate.org/resources/noaas -sea-level-rise-planning-tool; Sea Level Rise Planning Tool–New Jersey and New York State, last modified June 16, 2015, http://www.arcgis.com/home /item.html?id=2960f1e066544582ae0f0d988ccb3d27
22. Emanuel, *Divine Wind*, 149.
23. Ibid., 256; Kerry Emanuel, "The Hurricane–Climate Connection," in *Bulletin of the American Meteorological Society*, May 2008.
24. Morris A. Bender, Thomas R. Knutson, Robert E. Tuleya, Joseph J. Sirutis, Gabriel A. Vecchi, Stephen T. Garner, and Isaac M. Held, "Modeled Impact of Anthropogenic Warming on the Frequency of Intense Atlantic Hurricanes," in *Science* 327 (2010): 454; Thomas R. Knutson, John L. McBride, Johnny Chan, Kerry Emanuel, Greg Holland, Chris Landsea, Isaac Held, James P. Kossin, A.K. Srivastava, and Masato Sugi, "Tropical Cyclones and Climate Change," in *Nature Geoscience* (2010).

25. Directive 2007/60/EC of the European Parliament and of the Council of October 23, 2007 on the assessment and management of flood risks, http://eur-lex.europa.eu/legal-content/EN/TXT/?uri=CELEX:32007L0060

26. National Research Council et al., *Mapping the Zone*, 93.

27. Belgium: http://geoportal.ibgebim.be/webgis/Overstroming_kaart.phtml; Ireland: http://maps.opw.ie/fhrm/viewer/; Denmark: http://miljoegis.mim.dk/cbkort?profile=miljoegis_oversvoemmelsesdirektiv

28. France: http://www.aquitaine.developpement-durable.gouv.fr/cartes-et-rapports-d-accompagnement-des-tri-r912.html; "Member States' Examples of Flood Hazard and Flood Risk Maps," document prepared for the 2015 EU Water Conference, http://ec.europa.eu/environment/water/flood_risk/pdf/MS%20examples.pdf

29. For more about global climate models, see Paul Edwards, *A Vast Machine: Computer Models, Climate Data, and the Politics of Global Warming* (Cambridge, MA: MIT Press, 2010); and G. Flato, J. Marotzke, B. Abiodun, P. Braconnot, S.C. Chou, W. Collins, P. Cox, F. Driouech, S. Emori, V. Eyring, C. Forest, P. Gleckler, E. Guilyardi, C. Jakob, V. Kattsov, C. Reason, and M. Rummukainen, "2013: Evaluation of Climate Models," in *Climate Change 2013: The Physical Science Basis*, Contribution of Working Group I to the Fifth Assessment Report of the Intergovernmental Panel on Climate Change, ed. T.F. Stocker, D. Qin, G.-K. Plattner, M. Tignor, S.K. Allen, J. Boschung, A. Nauels, Y. Xia, V. Bex, and P.M. Midgley (Cambridge: Cambridge University Press, 2013), https://www.ipcc.ch/pdf/assessment-report/ar5/wg1/WG1AR5_Chapter09_FINAL.pdf

30. Ning Lin, Kerry Emanuel, Michael Oppenheimer, and Erik Vanmarcke, "Physically-Based Assessment of Hurricane Surge Threat under Climate Change," in *Nature Climate Change* 2:6 (2012): 462–67; Ning Lin, Kerry Emanuel, J.A. Smith, and Erick Vanmarcke, "Risk Assessment of Hurricane Storm Surge for New York City," in *Journal of Geophysical Research* 115 (2010).

31. Kerry Emmanuel, Ragoth Sundararajan, and John Williams, "Hurricanes and Global Warming: Results from Downscaling IPCC AR4 Simulations," in *Bulletin of the American Meteorological Society*, March 2008: 349; method also explained in Kerry Emanuel, Sai Ravela, Emanuel Vivant, and Camille Risi, "A Statistical Deterministic Approach to Hurricane Risk Assessment," in *Bulletin of the American Meteorological Society*, March 2006.

32. For more information about comparing climate models, see *2013: Evaluation of Climate Models*.

33. The other RCPs in the IPCC report are titled RCP2.6, RCP4.5, and RCP6.0. In contrast to RCP8.5, these pathways account for greenhouse gas emissions to decline at various points in the twenty-first century.

34. Richard A. Luettich and J.J. Westerink, "A Three Dimensional Circulation Model Using a Direct Stress Solution over the Vertical," in *Computational Methods in Water Resources IX, Volume 2: Mathematical Modeling in Water Resources*, ed. T. Russell et al. (Southampton, UK: Computational Mechanics Publications, 1992).

35. Lin et al., "Physically-Based Assessment," 5.

36. Robert E. Kopp, Radley M. Horton, Christopher M. Little, Jerry X. Mitrovica, Michael Oppenheimer, D.J. Rasmussen, Benjamin J. Strauss, and Claudia Tebaldi, "Probabilistic 21st and 22nd Century Sea-Level Projections at a Global Network of Tide-Gauge Sites," in *Earth's Future* 2 (2014): 383–406.

37. Ibid., 383–4.

38. Jack Eggleston and Jason Pope, 2013, "Land Subsidence and Relative Sea-Level Rise in the Southern Chesapeake Bay Region," in U.S. Geological Survey Circular 1392, 30 pp., http://dx.doi.org/10.3133/cir1392

39. In 1968, the Structural Engineers Association of California (SEAOC) recommended requirements for seismic design in the case of minor, moderate, and major earthquakes. This range is a precursor to the multiple-scenario planning of PBD.

40. "Next-Generation Performance-Based Seismic Design Guidelines: Program Plan for New and Existing Buildings," FEMA 445, August 2006, 3.

41. "Performance Based Seismic Design of Buildings: An Action Plan for Future Studies," FEMA 283, September 1996, 1.

42. "Next-Generation Performance," 2.

43. Ibid.

44. John King, "How Safe Are Rising S.F. Towers in the Wake of Napa Earthquake?" in *San Francisco Chronicle*, August 17, 2014.

Chapter 5

1. Jodi Wright-Gidley and Jennifer Marines, *Galveston: A City on Stilts* (Charleston, SC; Chicago, IL: Portsmouth, NH; San Francisco, CA: Arcadia Publishing, 2008).

2. For the history of the discovery, see Howard A. Kelly, *Walter Reed and Yellow Fever* (Baltimore, MD: The Medical Standards Book Company Publishers, 1906).

3. Henry Clay Weeks, ed. *Proceedings of the First General Convention to Consider the Questions Involved in Mosquito Extermination* (Brooklyn, NY: Eagle Book Printing Department, 1904).

4. Thomas J. Headlee, *Bulletin 348: The Mosquitoes of New Jersey and Their Control* (New Brunswick: New Jersey Agricultural Experiment Stations, 1921).

5. Nicholas Dagen Bloom and Mathew Gordon Lasner, eds., *Affordable Housing in New York: The People, Places, and Policies That Transformed a City* (Princeton, NJ: Princeton University Press, 2016), 83.

6. Jonathan Mahler, "How the Coastline Became a Place to Put the Poor," in *New York Times*, December 3, 2012.

Glossary

Adaptation: Adjustment in natural or human systems in anticipation of or in response to a changing environment in a way that effectively uses beneficial opportunities or reduces negative effects. (Executive Order 13653, 2013)

Adaptive management: A systematic approach to resource management that explores alternative ways to meet management objectives through prediction, implementation, monitoring, and updating knowledge, and subsequently adjusting management actions. (Department of the Interior)

Advanced Circulation Model (ADCIRC): ADCIRC is a hydrodynamic digital modeling technology that conducts short- and long-term simulations of tide and storm surge elevations and velocities in deep ocean, continental shelves, coastal seas, and small-scale estuarine systems. It is certified by the Federal Emergency Management Agency (FEMA) for use in performing storm surge analyses. (USACE)

Alluvial plain: A plain bordering a river, formed by the deposition of material eroded from areas of higher elevation and transported by the river to points along the floodplain. Alluvium is the material deposited by the stream, consisting of sand, silt, and other sediments derived from rocks. (USACE)

Astronomical tide: The tide levels and character that would result from the gravitational effects of the earth, sun, and moon without any atmospheric influences.

Attenuation: The loss or dissipation of wave energy, resulting in a reduction of wave height or amplitude.

Atoll terrace: An emergent, narrow, double-sloped ridge formed at the subtidal perimeter of a salt marsh footprint. Constructed of strategically placed and planted dredged material, this ridge or terrace evokes the appearance of a discontinuous atoll perimeter and functions as a sediment trap to support marsh accretion through sediment deposition. (SCR Jamaica Bay)

Base flood: The national standard for floodplain management is the 1 percent chance flood, known as the base flood. This flood has at least one chance in 100 of occurring in any given year. It is also called a 100-year flood. (USACE)

Base flood elevation (BFE): The elevation of the flood that has a 1 percent chance of occurring in any given year. (USACE)

Bathymetry: The measurement of the depths of water below sea level in oceans, seas, and lakes. (NOAA)

Bay nourishment: The introduction and strategic placement of material, generally quantities of sand or dredged material, by mechanical means within an estuarine embayment to supplement the natural processes of sediment delivery from the ocean via overwash and from the upland watershed. Reasons for nourishing a bay include creating reserves of sedimentary material and supplementing the cycles of sedimentary deposition at salt marsh islands within the intertidal zone, thus equalizing the rate of marsh accretion with the rate of sea level rise. (SCR)

Beach nourishment: The introduction of material, generally quantities of sand, by mechanical means along a shoreline to supplement the natural littoral drift. Reasons for nourishing a shoreline include controlling erosive forces, supplementing littoral drift to offset particular actions or works, and replenishing reserves of littoral material normally available in sand dunes. (USACE)

Beneficial use of dredged material: The use of dredged sediments as resource materials in productive ways to provide environmental, economic, or social benefit. Beneficial uses include agricultural uses, engineered uses, and environmental enhancement. The U.S. Army Corps of Engineers fully supports and strives to beneficially use dredged material in all circumstances where it is practical and cost-effective and where those beneficial uses can be accomplished in compliance with all requirements of federal law. (USACE)

Benthic: Pertaining to the subaquatic bottom or organisms that live on the bottom of water bodies. (USACE)

Berm: Narrow shelf of ground left naturally occurring or purposefully constructed at the base of a levee or offshore along the coast. (USACE)

Bulkhead: A structure or partition to retain or prevent sliding of the land. A secondary purpose is to protect the upland against damage from wave action. (USACE)

Check valve: A device that allows water to stream through a weir, outfall structure, or other conveyance in one direction but closes to prohibit backward flow of liquid.

Chenier ridge: A long, narrow wooded beach ridge or sandy hummock forming roughly parallel to a prograding shore, usually seaward of marsh and mudflat deposits (as along the southwest coast of Louisiana). (USACE)

Climate change: A nonrandom change in climate that is measured over several decades or longer. The change may be due to natural or human-induced causes. Measured change may include any long-term trend in mean sea level, wave height, wind speed, drift rate, and so on. (NOAA) The United Nations Framework Convention on Climate Change emphasizes the attribution of human activity, defining climate change as a change of climate that is attributed directly or indirectly to human activity that alters the composition of the global atmosphere and is in addition to natural climate variability observed over comparable time periods. (UNFCCC)

Coastal Storm Risk Management (CSRM): A U.S. Army Corps of Engineers planning process for applying policy to practice. CSRM addresses a life cycle approach and risk-informed decision making. The *CSRM Manual*, a framework for economic analyses of coastal projects, describes basic coastal processes and coastal engineering principles and models used in evaluating storm risk and long-term erosion and presents a discussion of National Economic Development benefits and costs as they relate to coastal storm risk management. (USACE)

Combined sewer overflow (CSO): Combined sewer systems are sewers that are designed to collect rainwater runoff, domestic sewage, and industrial wastewater in the same pipe. The combined sewer system generally transports its wastewater to a sewage treatment plant, but during periods of heavy rainfall or snowmelt, the wastewater volume can exceed the capacity of the treatment plant or the system itself. The system is designed to allow overflow in this situation, discharging untreated wastewater directly to streams, rivers, or other water bodies; these overflows are called combined sewer overflows (CSOs). (EPA)

Cut and fill: In earthmoving, a cut-and-fill operation is a procedure in which the elevation of a landform surface is modified by the removal or addition of surface material. This process may be human-made, through mechanical excavation, or natural, though water- or wind-driven erosion and deposition.

Delta: An alluvial deposit, usually triangular or semicircular, at the mouth of a river or stream. The delta is normally built up only where there is no tidal or current action capable of removing the sediment at the same rate as it is deposited, and hence the delta builds forward from the coastline. (USACE)

Digital elevation model (DEM): The representation of continuous elevation values over a topographic surface by a regular array of z-values, referenced to a common datum.

Dredged material: Material excavated from freshwater, estuarine, or marine waters. Once termed "spoils," suggesting equivalence with waste, dredged material is now considered a valuable resource for beneficial use, including agricultural uses, engineered uses, and environmental enhancement. (USACE)

Dune: A topographic ridge or mound form characterized by a steeper face on the side toward which it advances and a more gradual slope upwind or upcurrent. Commonly, dunes are composed of sand, but in some cases they consist of finer sediment. Dunes that have ceased moving are regarded as fixed and may be covered with vegetation and exhibit soil development.

Engineer Research and Development Center (ERDC): The mission of the U.S. Army Corp of Engineers' Engineer Research and Development Center is to help solve the nation's most challenging problems in civil and military engineering, geospatial sciences, water resources, and environmental sciences for the Army, Department of Defense, civilian agencies, and the nation's public good. ERDC was established on October 1, 1998 as an umbrella organization overseeing seven research laboratories in Illinois, Mississippi, New Hampshire, and Virginia, functioning as integrated teams of engineers and scientists to address a broad range of science and technology issues. (USACE)

Erosion: The wearing away of land by the action of natural forces such as currents, waves, wind, ice, and other forces. At the coast, the carrying away of material by wave action, tidal currents, or deflation.

Estuary: An embayment of the coast in which fresh river water entering at its head mixes with the saline ocean water. When tidal action is the dominant mixing agent it is usually called a tidal estuary. (NOAA)

Federal Emergency Management Agency (FEMA): FEMA is an agency of the U.S. Department of Homeland Security. Its mission is to support U.S. citizens and first responders to ensure that the nation works to build, sustain, and improve capability to prepare for, protect against, respond to, recover from, and mitigate all hazards. FEMA was initially created by the Presidential Reorganization Plan No. 3 of 1978 and implemented by two executive orders under President Jimmy Carter on April 1, 1979. (FEMA)

Fetch: The unobstructed distance over a bay or other body of water in which waves are generated by wind of relatively constant direction and speed. Wind fetch exerts energy on waves, causing them to be higher and more forceful upon impact with shorelines. (USACE)

Floodgate: A gravity outlet fitted with vertically hinged doors, opening if the inner water level is higher than the outer water level, so that drainage takes place during low water. (USACE)

Flood hazards of special concern: Flood-related hazards that are less common, more destructive, and harder to map than riverine, coastal, alluvial fan, and shallow flooding. Special hazards include coastal erosion, tsunamis, closed basin lakes, uncertain flow paths, dam breaks, ice jams, and mudflows. (FEMA)

Flood Insurance Rate Map (FIRM): The official map of a community on which the Federal Emergency Management Agency (FEMA) has delineated both the special hazard areas and the risk premium zones applicable to the community. (FEMA)

Floodplain: A flat tract of land bordering a river, mainly in its lower reaches, and consisting of alluvium deposited by the river. It is formed by the sweeping of the meander belts downstream, thus widening the valley, the sides of which may be some kilometers apart. During floods, when the river overflows its banks, sediment is deposited along the valley banks and plains. (USACE)

Floodplain management: Floodplain management is a decision-making process that aims to achieve the wise use of the nation's floodplains. "Wise use" means both reduced flood losses and protection of the natural resources and function of floodplains. (FEMA)

Flood risk: A flood inundates a floodplain. Most floods fall into three major categories: riverine flooding, coastal flooding, and shallow flooding. Alluvial fan flooding is another type of flooding more common in the mountainous western states. Flood risk is the likelihood and adverse consequences of flooding. Flood risk for assets and people at any location in a floodplain is a function of flood hazard at that location and their exposure and vulnerability to the flood hazard. In areas served by flood hazard reduction infrastructure, the remaining risk is often called *residual risk*. (USACE/FEMA)

Flushing tunnel: A below-grade infrastructural tunnel with a mechanized pumping system for improving water quality, reducing residence time, and encouraging hydrologic flow between bay and ocean. (SCR Jamaica Bay)

Freeboard: At a given time, the vertical distance between the water level and the top of a structure. For regulatory and design purposes, the additional height of a structure above the design high water level to prevent overflow. (USACE/NOAA)

Global climate model: General circulation models, also known as a global climate models, are numerical models or mathematical formulations of the processes that make up the climate system, including the physical processes in the atmosphere, ocean, cryosphere, and land surface. These are the most advanced tools currently available for simulating the response of the global climate system to increasing greenhouse gas concentrations. (IPCC)

Hindcast: The determination through empirical relations or numerical models of wave heights, periods, directions, and such factors as storm surge from historical weather charts or other historical records. (USACE)

Horizontal datum: A geodetic datum for any extensive measurement system of positions, usually expressed as latitude–longitude coordinates, on the earth's surface. A horizontal geodetic datum may be local or geocentric. (ESRI)

Hummock: A hillock, knoll, or mound; a piece of forested ground rising above a marsh.

Hurricane: An intense tropical cyclone in the Atlantic, Caribbean Sea, Gulf of Mexico, or eastern Pacific in which winds tend to spiral inward toward a cone of low pressure, with maximum surface wind velocities that equal or exceed 74 miles per hour (64 knots) for several minutes or longer at some points. (USACE/NOAA)

Hydrodynamics: Synonymous with fluid dynamics. In physics and engineering, fluid dynamics is a subdiscipline of fluid mechanics that describes the flow of fluids.

Hydrology: Hydrology is the science that encompasses the occurrence, distribution, movement, and properties of the waters of the earth and their relationship with the environment in each phase of the hydrologic cycle. The water cycle, or hydrologic cycle, is a continuous process by which water is purified by evaporation and transported from the earth's surface (including the oceans) to the atmosphere and back to the land and oceans. (NOAA/USGS)

Island motor: A process of strategic sediment distribution that harnesses the natural forces of tides, current, and wind to encourage sediment capture and deposition in a desired location to support salt marsh island accretion. This sediment source may be beneficially placed dredged material or naturally occurring sediment suspended within the water column. (SCR Jamaica Bay)

Jetty: On open seacoasts, a structure extending into a body of water, designed to prevent shoaling of a channel by littoral materials and to direct and confine the stream or tidal flow. Jetties are built at the mouth of a river or entrance to a bay to help deepen and stabilize a channel and facilitate navigation. (USACE)

Jurisdiction of the U.S. Army Corps of Engineers: Section 404 of the Clean Water Act defines the landward limit of the USACE jurisdiction as the high tide line in tidal waters and the ordinary high water mark as the limit in nontidal waters. When adjacent wetlands are present, the limit of jurisdiction extends to the limit of the wetland. (USACE)

Levee: Earthen structure (e.g., dike or embankment) built to contain periodic floodwater from river systems to within a specific area of the floodplain. (USACE)

Littoral drift: The sedimentary material moved or transported in the littoral zone parallel to the shoreline under the influence of waves and currents.

Maritime forest: Coastal forest communities within range of salt spray or mist typically dominated by closed canopies, located on the mainland side of a barrier island. The biomass of these forests, when densely planted, can attenuate waves and storm surge, thus reducing inundation of coastal communities.

National Flood Insurance Program (NFIP): NFIP is a program created by the U.S. Congress in 1968 through the National Flood Insurance Act of 1968 (Public Law 90-448). The program, administered by the government, enables property owners in participating communities to purchase insurance protection against losses from flooding and requires flood insurance for all loans or lines of credit that are secured by existing buildings, manufactured homes, or buildings under construction that are located in a community that participates in the NFIP. (FEMA)

National Oceanic and Atmospheric Administration (NOAA): NOAA is an American scientific agency within the U.S. Department of Commerce that focuses on the conditions of the oceans and the atmosphere. NOAA warns of dangerous weather, charts seas, guides the use and protection of ocean and coastal resources, and conducts research to provide understanding and improve stewardship of the environment. (NOAA)

Natural and nature-based features (NNBF): Natural features are created and evolve over time through the actions of physical, biological, geologic, and chemical processes operating in nature. Natural coastal features take a variety of forms, including reefs, barrier islands, dunes, beaches, wetlands, and maritime forests. Nature-based features are those that may mimic characteristics of natural features but are created by human design, engineering, and construction to provide specific services such as coastal storm risk management. (USACE)

Nonstructural flood risk management measures: Nonstructural flood risk management measures are proven methods and techniques for reducing flood risk and flood damages incurred within floodplains. Techniques include elevation of structures, relocation of structures, acquisition of structures and the land, wet and dry floodproofing, small-scale berms and floodwalls, flood warning systems, flood emergency preparedness plans, and land use regulations. (USACE)

North American Datum of 1983 (NAD 83): The horizontal control datum for the United States, Canada, Mexico, and Central America, based on a geocentric origin and the Geodetic Reference System 1980. NAD 83 is the current geodetic reference system. NAD 83 is based on the adjustment of 250,000 points, including 600 satellite Doppler stations, that constrain the system to a geocentric origin. (NOAA)

North American Vertical Datum of 1988 (NAVD 88): A fixed reference for elevations determined by geodetic leveling. The datum was derived from a general adjustment of the first-order terrestrial leveling nets of the United States, Canada, and Mexico. In the adjustment, only the height of the primary tidal benchmark, referenced to the International Great Lakes Datum of 1985 (IGLD 85) local mean sea level height value, at Father Point, Rimouski, Quebec, Canada, was held fixed, thus providing the minimum constraint. (NOAA)

Overtopping: Passing of water over the top of a structure, as in wave runup or surge action.

Overwash plain: A defined zone of low topographic elevation along a barrier island or peninsula across which water and suspended sediment may flow during extreme water level fluctuations. The zone of the overwash plain is selected by its existing low terrain, and the inundation zone is limited through the construction of adjacent elevated berms. Overwash plains allow sediment from the ocean to nourish the back-bay system and provide additional outlets for back-bay floodwaters to return to the ocean. (SCR Jamaica Bay)

Regional sediment management (RSM): Developed by the U.S. Army Corps of Engineers, RSM is a method that manages sediment to benefit a region. This method allows the use of natural processes, improves the environment, and potentially saves money. RSM includes the entire environment from the watershed to the sea, accounts for the effects of human activities on sediment erosion and transport, and protects and enhances the nation's natural resources while balancing national security and economic needs. (USACE)

Residence time: A measure of the average time a physical, chemical, or biological substance spends within a physical system or reservoir (e.g., atmosphere, oceans, soil). In the case of the coastal ocean, a measure of residence time can be extremely useful in determining transport and fate of contaminants and organisms in estuarine systems. (NOAA)

Resilience: The U.S. Army Corps of Engineers (per Executive Order 13653, 2013) defines resilience as the ability to anticipate, prepare for, respond to, and adapt to changing conditions and to withstand and recover rapidly from disruptions with minimal damage. The ecological definition of resilience is the capacity of an ecosystem to respond to a perturbation or disturbance of stochastic nature by resisting damage and recovering quickly. (USACE/Holling, 2002)

Revetment: Retaining wall or facing of stone, concrete, or other material, created to protect an embankment or shore structure against erosion by wave action or currents. (USACE)

Risk: The potential for realization of unwanted, adverse consequences; estimation of risk is usually based on the expected result of the conditional probability of the occurrence of event multiplied by the consequence of the event when it occurs. (USACE)

River diversion: An artificial channel that is used to divert all or a portion of the flow of a river from its natural or primary course, often used to divert freshwater from a main channel to a different distributary so that water and sediment are distributed across a broader range of the river delta.

Salt marsh: A marsh periodically flooded by saltwater. The primary herbaceous salt marsh species on the east coast of the United States is *Spartina alterniflora*, smooth cordgrass, which thrives in the intertidal zone from mean tide level to mean high water. (USACE)

Sand motor: An experiment in the management of dynamic coastlines, the sand motor (sometimes called the sand engine) is located offshore of Ter Heijde, Zuid-Holland in the Netherlands. The sand motor is a large artificially placed sand bank distributed by wind and waves to increase coastal protection and create a dynamic natural and recreational area. (Rijkswaterstaat)

Sea, Lake, and Overland Surges from Hurricanes (SLOSH): The SLOSH model is a computerized numerical model developed by the National Weather Service to estimate storm surge heights resulting from historical, hypothetical, or predicted hurricanes by taking into account the atmospheric pressure, size, forward speed, and track data. These parameters are used to create a model of the wind field that drives the storm surge. (NOAA)

Sea level rise: An increase in the volume of water in the world's oceans, resulting in an increase in global mean sea level. Sea level rise is usually attributed to global climate change by the thermal expansion caused by the warming of the water in oceans and increased melting of land-based ice such as glaciers and ice sheets. (NOAA)

Sediment: Material, such as sand, silt, or clay, suspended in or settled on the bottom of a water body. Sediment input to a body of water comes from natural sources, such as erosion of soils and weathering of rock, or as the result of anthropogenic activities such as forest or agricultural practices, or construction activities. The term *dredged material* refers to material that has been dredged from a water body, whereas the term *sediment* refers to material in a water body before the dredging process. (USACE)

Sewershed: Defined as an analogy to the concept of a natural watershed, which refers to an area draining to a single point in a stream network, a sewershed is a catchment area defined by storm and sewage drain infrastructure emptying into a common outlet. Sewage or stormwater flows to a single point connection at a sewer interceptor pipe and onward to a single pump station or treatment plant. With a combined sewage and stormwater system, local water treatment plants are often overwhelmed, and stormwater and untreated sewage are discharged through outfall pipes to streams.

Shoreline: The intersection of a specified plane of water with the shore or beach. The line delineating the shoreline on NOAA's National Ocean service nautical charts and surveys approximates the mean high-water line. (USACE)

Storm surge: The abnormal rise of water generated by a storm or hurricane due to a combination of wind and low atmospheric pressure, over and above the normal astronomical tide. It is expressed in terms of height above predicted or expected tide levels. (NOAA)

Storm tide: The water level due to the combination of storm surge and the astronomical tide, expressed in terms of height above a vertical or tidal datum. (NOAA)

Subsidence: Sinking or downwarping of a part of the earth's surface.

Tidal marsh inlet: A long and narrow indentation of a shoreline connecting the ocean through salt marsh to the bay during high tide conditions.

Topobathy: Topographic and bathymetric elevation data merged in reference to a common vertical datum. (NOAA)

Topography: The measurements of the heights of the earth's terrain and relief surface features above sea level. (NOAA)

Tropical cyclone: A warm-core, nonfrontal synoptic-scale cyclone, originating over tropical or subtropical waters with organized deep convection and a closed surface wind circulation about a well-defined center. (NOAA)

U.S. Army Corps of Engineers (USACE): USACE is a U.S. federal agency under the jurisdiction of the Department of Defense and one of the world's largest public engineering, design, and construction management agencies. Missions include civil engineering of locks, dams, maintenance dredging, and flood control as well as environmental regulation and ecosystem restoration. (USACE)

U.S. Geological Survey (USGS): The USGS is a U.S. federal scientific agency under the jurisdiction of the U.S. Department of the Interior, studying the United States' natural resources, natural hazards, ecosystems, and environmental health and the impacts of climate and land use change. Science disciplines of the agency include biology, geography, geology, and hydrology.

Verge enhancement: The strategic gradient enhancement and elevation of coastal edges and margins for protection from waves, storms, and erosion through a layered buffering system of marsh terraces, earthen berms, and attenuation forests. These elevated gradient lands may connect with existing linear infrastructures such as roadbeds, railbeds, and bridges and may also be used for accessible public parklands. (SCR Jamaica Bay)

Vertical datum: A base elevation used as a reference from which to measure heights (or depths). Similarly, a tidal datum is a base elevation defined by a certain phase of the tide. (NOAA)

Vulnerability: The U.S. Army Corps of Engineers defines *vulnerability* as a function of the hazard to which a system is exposed, the sensitivity of the system to the hazard, and the system's adaptive capacity. (USACE)

Water quality: A measure of the suitability of water for a particular use by one or more biotic species based on selected physical, chemical, biological, and radiological characteristics. (USGS)

Watershed: An area of land that drains all the streams and rainfall to a common outlet such as the outflow of a reservoir, mouth of a bay, or any point along a stream channel; a drainage basin or catchment area. (USGS)

Waterways Experiment Station (WES): A 673-acre complex in Vicksburg, Mississippi, built in 1930 as a U.S. Army Corps of Engineers federal hydraulics research facility. It is the site of the Engineer Research and Development Center (ERDC) headquarters and includes the following research and development laboratories: Coastal and Hydraulics, Geotechnical and Structures, Environmental, and Information Technology. WES was established in response to the Great Mississippi Flood of 1927. (USACE)

Wave: Change in the elevation of water in the ocean caused by the motion of currents and wind action. (NOAA)

Wetland: Land whose saturation with water is the dominant factor in determining the nature of soil development and the types of plant and animal communities that live in the soil and on its surface. In hydrologic terms, an area that is regularly wet or flooded and has a water table that stands at or above the land surface for at least part of the year. (USACE/NOAA)

Index

Note: Color plates are indicated by the letter p followed by the plate number. Figures are indicated by a page number followed by the letter f.